Starrise and Starset Guide

A Guide for Amateur Observers

by STAN POLCZYK

IN APPRECIATION

to my dad
who showed me the Milky Way
while I was still a young child,
and to my relatives and friends
who encouraged and inspired
my interest in astronomy.

SPECIAL THANKS

to my relatives, friends and associates
who assisted with the timetables,
the proofreading and who in any way
helped get the manuscript to press.

Library of Congress Catalogue Card Number 90-091638

ISBN 0-9626004-0-7

Photograph, maps and design by author.

Published by Stan Polczyk
P.O. Box 384, Keasbey, New Jersey 08832

Printed in the United States of America
by American Press, Wall, New Jersey.

THE JOY OF ASTRONOMY

IS TO LOOK AT,

OR AT LEAST

TO IMAGINE TO LOOK AT,

THE STARRY SKY.

"A Voyage from Starrise to Starset"
--the cover photo

Near the hour of midnight, the midnight of April 11-12, 1986, the author was in southern Florida to view Halley's Comet.

The wind carried the clouds and the haze away so that stars close to the horizon could be seen.

The trails captured in the unique photograph include Halley's Comet, Alpha and Beta Centauri rising as well as the Southern Cross setting.

The long line across the picture was caused by the lights of an uninvited airplane.

The shorter line, centered at the bottom of the picture, marks the horizon. This line was caused by the lights of a boat. The boat's voyage, which looks as if it began near Alpha in Centaurus and ended near Alpha in the Southern Cross, took about fifteen minutes.

CONTENTS

Statement of Purpose

Some stars, known as circumpolar stars, always remain above the horizon. Some stars always remain below the horizon and are never visible. Meanwhile, some stars rise and set. As is evident from the title, the information in this guide is about the third group, the stars that rise and set.

Many books explain the rising and setting of stars, but there are only a few resources available to help star observers who prefer not to use advanced mathematics, electronic calculators and computers, but who still wish to make some reasonable estimate of the time of starrise and the time of starset.

The purpose of this **Starrise and Starset Guide** is to help star observers estimate the time of starrise and the time of starset. For advanced applications and advanced study, observers must calculate the moment of rising and setting more precisely. However, since stars are best viewed at some distance above the horizon, star observers who estimate the moment of rising and setting usually are not at a disadvantage when it comes to enjoying the sky.

The proceedures described in this book may be followed only when the exact minute, second or part of a second is not required. These time estimates are **not** intended for specialized applications such as religious and professional astronomy, formal academic study, navigation and aviation.

Asking the Question . . .

I developed an interest in the phenomena of starrise and starset during my younger days.

Even before I reached the library, I had an idea about what I wanted to look for. I knew that, with the help of a rotating star chart (a planisphere), I could estimate the time a star would be at the meridian; that is, halfway between the time of rising and the time of setting. Now, I had to find the timetables which would tell me the number of hours and minutes to subtract to find the time of starrise and to add to find the time of starset!

"Looks easy!" I thought!

Not so! Instead of a timetable, I was presented with mathematical formulas which, I now understand, were derived from spherical trigonometry! With great effort, I tried one or two examples. However, the language was too technical and the math was too advanced. This experience caused me to delay my efforts for a while.

Through the years, I learned more about the mathematical formulas used in astronomy. I also developed a better appreciation of the information I searched for many years ago, which I use as a convenient shortcut while planning to observe the sky. The information is presented in this **Starrise and Starset Guide** so that other star observers may locate this information easily and benefit from it.

--the Author

. . . and Applying the Answer

Many star observers, experienced as well as beginners, use a simple mechanical device (a rotating star chart, often called a planisphere) to estimate the time a star will reach the meridian. In the case of stars that rise and set, the meridian is located halfway between starrise and starset.

Since the time of starrise and the time of starset is not the same for every star, the timetables in this **Starrise and Starset Guide** provide the number of hours and minutes to be subtracted (from the time a star is on the meridian) to find the estimated time of starrise and to be added (to the time a star is on the meridian) to find the estimated time of starset.

An estimated time of starrise as well as an estimated time of starset may be used as guideposts for planning visual observations, because stars, when seen in nature, are seldom observed and studied exactly at starrise or exactly at starset. Atmospheric haze, clouds, trees, hills and mountains as well as lights and buildings often produce poor conditions for observing stars near the horizon. Frequently, the stars cannot be seen after starrise and before starset for a number of minutes, an hour or even longer.

When we are prepared to observe a star from the estimated time of its rising to the estimated time of its setting, our efforts are rewarded when nature provides good local conditions for observing stars. The author, for example, photographed Alpha and Beta Centauri rising as well as the Southern Cross setting from a location in southern Florida after the wind carried away the haze and the clouds.

The photograph printed on the cover of this **Starrise and Starset Guide**, which includes Halley's Comet, reminds us of stars (and other objects) which we hope to observe during their east-to-west journeys across the sky. The star Alpha in the Southern Cross, the most distant star from the circumpolar region at that location, rises about one hour and thirty-five minutes before and sets about one hour and thirty-five minutes after it reaches the meridian. In contrast, Halley's Comet, as photographed, rises about four hours before and sets about four hours after it reaches the meridian.

If the photograph would have included all stars that rise and set which were visible at the location, we would find that the number of hours would increase for stars closer to the circumpolar region. (The time would vary proportionately from declination to declination and from latitude to latitude.) The list of times for stars in the extended photograph would resemble the list in the timetables in this guide for that latitude.

Whenever we estimate the time of starrise and the time of starset, our thoughts should be focused on the stars and on other objects in the sky. The estimated time of starrise marks the start of their daily east-to-west journeys across the sky. The estimated time of starset marks the end of those journeys. Hopefully, nature will permit our eyes to witness those journeys.

About Atmospheric Refraction

As starlight enters into and travels through the earth's atmosphere, the starlight itself is bent and the star's position in the sky appears shifted. For most of the sky, the shift is relatively small. There is no shift at the zenith. The greatest shift is at the horizon. Since this phenomenon, known as atmospheric refraction, causes the stars to appear closer to the zenith (directly overhead), stars at the horizon seem to rise earlier and set later.

On the one hand, the additional time a star may spend above the horizon because of atmospheric refraction is a welcome bonus--nature's way of helping star observers catch a better glimpse of stars that remain close to the horizon as they travel across the sky!

On the other hand, the mathematical equations do not tell the entire story. The time gained from atmospheric refraction is not distributed evenly during the journey across the sky. Paradoxically, atmospheric conditions hinder visual observations near the horizon.

Before starrise and after starset, as predicted without time adjustments, star observers probably will encounter poor conditions for visual observations.

At or near starrise and starset, as predicted without time adjustments, star observers will notice that starrise has taken place a little earlier and that starset will take place a little later because of atmospheric refraction and other phenomena.

Between starrise and starset, as predicted without time adjustments, star observers need not employ time adjustments to enjoy the sky.

THE FIVE STEPS

The Planisphere
(Rotating Star Chart)

* The planisphere, marketed under a variety of names, is a mechanical calculator which can be used by star observers, experienced as well as beginners, to estimate the time of day when a specific star will be located halfway between rising and setting.

* Before you begin you may wish to add a string or thread running directly north-south on your planisphere (rotating star chart).

* * * *

A. Rotate the planisphere until the star you wish to observe is on the meridian (the north-south line in the center of the chart), which is halfway between starrise and starset.

B. Find the month and date along the edge of your planisphere.

C. Read the time that is next to the month and date.

D. Add an hour for daylight savings time, etc., when applicable.

* * * *

* If the star you are looking for is not on your planisphere, you may still find the time by using a star directly north or directly south of the star you wish to observe. A star is north or south of another star if both stars have the same R.A.--right ascension. (Right ascension could be thought of as "longitude" in the sky.)

* For greater accuracy, you may wish to find out how far the observation site is east or west of the center of the time zone, and add or subtract the appropriate time correction.

STEP 1. **FIND THE ESTIMATED TIME**
****

THE STAR WILL BE ON THE MERIDIAN

ON A PARTICULAR DAY WITH THE HELP OF

A PLANISPHERE (a rotating star chart).

This information will be used in the final step.

For more details about the planisphere, see page 16.

Example: Alpha in the Southern Cross is the southernmost star in the cover photo. On April 11, this star will be halfway between starrise and starset at approximately 11:10 P.M.

(Since the star Alpha in the Southern Cross is not on our planisphere, the star Beta in Corvus, which is almost directly north, was used.)

STEP 2. **FIND THE LATITUDE**
****

OF THE OBSERVATION SITE

from a road map or an atlas.

Then, from this Starrise and Starset Guide **use the timetables for that latitude.**

Example: The cover photo was taken in southern Florida which is located at approximately 25° on the earth's northern hemisphere. We will use the timetables for latitude 25.

STEP 3. **FIND THE STAR'S DECLINATION**
****

("latitude" in the sky) **from a star map**

AND DETERMINE WHETHER TO USE

TIMETABLE A OR TIMETABLE B.

For more details about which timetable to use, turn to the timetables or to page 22.

Example: The declination of the star Alpha in the Southern Cross is almost -63°, which is 63° in the southern hemisphere of the sky. Since our observation site is in the earth's northern hemisphere, we will use Timetable B.

STEP 4. **FROM THE APPROPRIATE TIMETABLE,**
****

FIND THE TIME

FOR THE STAR'S DECLINATION.

Example: For a star located at 63°, the timetable we are using, Timetable B, reads: 1:35, which is 1 hour and 35 minutes.

STEP 5. MAKE YOUR CALCULATIONS.
****

5-A. To find THE TOTAL TIME A STAR IS VISIBLE, add the time found in the timetable (step 4) to itself.

Example: We add 1:35 (1 hr 35 min) to itself:

```
    1 hr  35 min
   +1 hr  35 min
   -------------
    3 hrs 10 min (3:10)
```

5-B. To find THE TIME OF STARRISE, subtract the time found in the timetable (step 4) from the time the star will be on the meridian (step 1).

Example: We subtract 1:35 (step 4) from 11:10 P.M., April 11 (step 1):

```
    11 hrs 10 min
   - 1 hr  35 min
   --------------
     9 hrs 35 min (9:35 P.M., April 11).
```

5-C. To find THE TIME OF STARSET, add the time found in the timetable (step 4) to the time the star will be on the meridian (step 1).

Example: We add 1:35 (step 4) to 11:10 P.M., April 11 (step 1):

```
    11 hrs 10 min
   + 1 hr  35 min
   --------------
    12 hrs 45 min (12:45 A.M., April 12).
```

Thus, at our observation site's latitude, the star Alpha in the Southern Cross rises at approximately 9:35 P.M. on April 11 and sets at approximately 12:45 A.M. on April 12. The star is above the horizon for approximately 3 hours and 10 minutes.

Possible Voyages in Astronomy

You may use this **Starrise and Starset Guide**:

- --to plan observations of stars near the southern horizon from northern hemisphere locations.
- --to plan observations of stars near the northern horizon from southern hemisphere locations.
- --to plan for observations in your backyard; that is, at your own latitude.
- --to plan for observations north or south of your latitude.
- --to dream about future trips to observe the stars at other latitudes.
- --to take imaginary voyages to latitudes you probably will never travel to.
- --to take an imaginary journey from the south pole to the north pole to study the position of a specific star at various latitudes.

This **Starrise and Starset Guide** does not deal with the rising and setting of the sun, moon and planets directly, because this information is readily available in many other publications.

IMPORTANT CAUTION: If you plan to observe the nearest star, the sun, you must take adequate precautions to protect your eyes. LOOKING AT THE SUN DIRECTLY WILL CAUSE PERMANENT EYE INJURY AND PERMANENT BLINDNESS. Before attempting to use any special technique(s), be sure to read the recommendations of several reputable publications.

THE TIMETABLES

Which Timetable to Use?

Timetable A is for stars **in the same celestial hemisphere** as the circumpolar region.

Timetable B is for stars **in the celestial hemisphere opposite** the circumpolar region.

* * * *

Use **Timetable A:**

--when observing stars in the **northern** hemisphere of the sky from the earth's **northern** hemisphere.
--when observing stars in the **southern** hemisphere of the sky from the earth's **southern** hemisphere.

SUMMARY: NORTH from NORTH or SOUTH from SOUTH use Timetable A.

* * * *

Use **Timetable B:**

--when observing stars in the **southern** hemisphere of the sky from the earth's **northern** hemisphere.
--when observing stars in the **northern** hemisphere of the sky from the earth's **southern** hemisphere.

SUMMARY: SOUTH from NORTH or NORTH from SOUTH use Timetable B.

Working with the Timetables:

* Timetables derived from the basic formula are not new to mathematics and astronomy, but the usual presentation and application is much more technical. The timetables and the method described in this booklet may be used by everyone who knows how to add and subtract hours and minutes.

* The timetables are concerned primarily with stars that rise and set. The information is divided into Timetable A and Timetable B for convenience. Likewise, time in the timetables is simply called time (rather than by a technical name).

* The timetables take into account the star's location in the sky (the declination) and the observer's location on the planet (the latitude). To estimate the time of a star's rising and setting, information obtained from a planisphere (a rotating star chart) may be used to satisfy the remainder of the mathematical formula.

* Three reference points may be easily located in the timetables; they are:
    - **12:00**, the time for the declination at the edge of the circumpolar region,
    - **6:00**, usually the time for the declination at the celestial equator (0°), and
    - **0:00**, the time for the declination at the edge of the region of the sky which is never visible.

* The time is given in hours and completed minutes. Extra seconds are not rounded to the next minute.

* The timetables, like the basic formula, do not contain time adjustments for refraction and other atmospheric and local phenomena. (A time adjustment for the earth's motion is explained on page 62.)

TIMETABLE A--for stars in the **SAME CELESTIAL HEMISPHERE** as the circumpolar region.

Latitude	Declination	Time	Declination	Time	Declination	Time
90	1	12:00				
	0	0:00				
89	1	12:00				
	0	6:00				
88	2	12:00				
	1	7:59				
	0	6:00				
87	3	12:00				
	2	8:47				
	1	7:17				
	0	6:00				
86	4	12:00	0	6:00		
	3	9:14				
	2	7:59				
	1	6:57				
85	5	12:00	1	6:46		
	4	9:32	0	6:00		
	3	8:27				
	2	7:34				
84	6	12:00	2	7:17		
	5	9:45	1	6:38		
	4	8:46	0	6:00		
	3	7:59				
83	7	12:00	3	7:41		
	6	9:55	2	7:06		
	5	9:01	1	6:32		
	4	8:18	0	6:00		
82	8	12:00	4	7:59	0	6:00
	7	10:03	3	7:27		
	6	9:13	2	6:57		
	5	8:34	1	6:28		
81	9	12:00	5	8:14	1	6:25
	8	10:10	4	7:44	0	6:00
	7	9:23	3	7:17		
	6	8:46	2	6:50		
80	10	12:00	6	8:26	2	6:45
	9	10:15	5	7:58	1	6:22
	8	9:31	4	7:33	0	6:00
	7	8:56	3	7:09		

TIMETABLE B--for stars in the celestial hemisphere
OPPOSITE THE CIRCUMPOLAR REGION.

				Declination	Time	Latitude
				0	0:00	90
				1	0:00	89
				0	6:00	
				2	0:00	88
				1	4:00	
				0	6:00	
				3	0:00	87
				2	3:12	
				1	4:42	
				0	6:00	
		0	6:00	4	0:00	86
				3	2:45	
				2	4:00	
				1	5:02	
		1	5:13	5	0:00	85
		0	6:00	4	2:27	
				3	3:32	
				2	4:25	
		2	4:42	6	0:00	84
		1	5:21	5	2:14	
		0	6:00	4	3:13	
				3	4:00	
		3	4:18	7	0:00	83
		2	4:53	6	2:04	
		1	5:27	5	2:58	
		0	6:00	4	3:41	
0	6:00	4	4:00	8	0:00	82
		3	4:32	7	1:56	
		2	5:02	6	2:46	
		1	5:31	5	3:26	
1	5:34	5	3:45	9	0:00	81
0	6:00	4	4:15	8	1:49	
		3	4:42	7	2:36	
		2	5:09	6	3:13	
2	5:14	6	3:33	10	0:00	80
1	5:37	5	4:01	9	1:44	
0	6:00	4	4:26	8	2:28	
		3	4:50	7	3:03	

TIMETABLE A--for stars in the **SAME CELESTIAL HEMISPHERE** as the circumpolar region.

Latitude	Dec	Time	Dec	Time	Dec	Time	Dec	Time	Dec	Time
79	11	12:00	7	8:36	3	7:02				
	10	10:20	6	8:10	2	6:41				
	9	9:38	5	7:46	1	6:20				
	8	9:05	4	7:24	0	6:00				
78	12	12:00	8	8:45	4	7:16	0	6:00		
	11	10:24	7	8:21	3	6:57				
	10	9:44	6	7:58	2	6:37				
	9	9:12	5	7:37	1	6:18				
77	13	12:00	9	8:53	5	7:29	1	6:17		
	12	10:28	8	8:29	4	7:10	0	6:00		
	11	9:49	7	8:08	3	6:52				
	10	9:19	6	7:48	2	6:34				
76	14	12:00	10	9:00	6	7:39	2	6:32		
	13	10:31	9	8:37	5	7:22	1	6:16		
	12	9:53	8	8:17	4	7:05	0	6:00		
	11	9:24	7	7:58	3	6:48				
75	15	12:00	11	9:06	7	7:49	3	6:45		
	14	10:34	10	8:44	6	7:32	2	6:29		
	13	9:57	9	8:24	5	7:16	1	6:14		
	12	9:29	8	8:06	4	7:00	0	6:00		
74	16	12:00	12	9:11	8	7:57	4	6:56	0	6:00
	15	10:36	11	8:50	7	7:41	3	6:42		
	14	10:01	10	8:31	6	7:26	2	6:27		
	13	9:34	9	8:14	5	7:11	1	6:13		
73	17	12:00	13	9:16	9	8:04	5	7:06	1	6:13
	16	10:38	12	8:56	8	7:49	4	6:52	0	6:00
	15	10:04	11	8:37	7	7:34	3	6:39		
	14	9:38	10	8:20	6	7:20	2	6:26		
72	18	12:00	14	9:20	10	8:11	6	7:15	2	6:24
	17	10:40	13	9:01	9	7:56	5	7:02	1	6:12
	16	10:07	12	8:43	8	7:42	4	6:49	0	6:00
	15	9:42	11	8:26	7	7:28	3	6:37		
71	19	12:00	15	9:24	11	8:17	7	7:23	3	6:35
	18	10:42	14	9:05	10	8:03	6	7:11	2	6:23
	17	10:10	13	8:48	9	7:49	5	6:58	1	6:11
	16	9:45	12	8:32	8	7:36	4	6:46	0	6:00
70	20	12:00	15	9:09	10	7:55	5	6:55	0	6:00
	19	10:44	14	8:52	9	7:43	4	6:44		
	18	10:12	13	8:37	8	7:30	3	6:33		
	17	9:48	12	8:22	7	7:18	2	6:22		
	16	9:27	11	8:09	6	7:07	1	6:10		

TIMETABLE B--for stars in the celestial hemisphere OPPOSITE THE CIRCUMPOLAR REGION.

Dec	Time	Dec	Time	Dec	Time	Dec	Time	Dec	Time	Latitude
				3	4:57	7	3:23	11	0:00	79
				2	5:18	6	3:49	10	1:39	
				1	5:39	5	4:13	9	2:21	
				0	6:00	4	4:35	8	2:54	
		0	6:00	4	4:43	8	3:14	12	0:00	78
				3	5:02	7	3:38	11	1:35	
				2	5:22	6	4:01	10	2:15	
				1	5:41	5	4:22	9	2:47	
		1	5:42	5	4:30	9	3:06	13	0:00	77
		0	6:00	4	4:49	8	3:30	12	1:31	
				3	5:07	7	3:51	11	2:10	
				2	5:25	6	4:11	10	2:40	
		2	5:27	6	4:20	10	2:59	14	0:00	76
		1	5:43	5	4:37	9	3:22	13	1:28	
		0	6:00	4	4:54	8	3:42	12	2:06	
				3	5:11	7	4:01	11	2:35	
		3	5:14	7	4:10	11	2:53	15	0:00	75
		2	5:30	6	4:27	10	3:15	14	1:25	
		1	5:45	5	4:43	9	3:35	13	2:02	
		0	6:00	4	4:59	8	3:53	12	2:30	
0	6:00	4	5:03	8	4:02	12	2:48	16	0:00	74
		3	5:17	7	4:18	11	3:09	15	1:23	
		2	5:32	6	4:33	10	3:28	14	1:58	
		1	5:46	5	4:48	9	3:45	13	2:25	
1	5:46	5	4:53	9	3:55	13	2:43	17	0:00	73
0	6:00	4	5:07	8	4:10	12	3:03	16	1:21	
		3	5:20	7	4:25	11	3:22	15	1:55	
		2	5:33	6	4:39	10	3:39	14	2:21	
2	5:35	6	4:44	10	3:48	14	2:39	18	0:00	72
1	5:47	5	4:57	9	4:03	13	2:58	17	1:19	
0	6:00	4	5:10	8	4:17	12	3:16	16	1:52	
		3	5:22	7	4:31	11	3:33	15	2:17	
3	5:24	7	4:36	11	3:42	15	2:35	19	0:00	71
2	5:36	6	4:48	10	3:56	14	2:54	18	1:17	
1	5:48	5	5:01	9	4:10	13	3:11	17	1:49	
0	6:00	4	5:13	8	4:23	12	3:27	16	2:14	
0	6:00	5	5:04	10	4:04	15	2:50	20	0:00	70
		4	5:15	9	4:16	14	3:07	19	1:15	
		3	5:26	8	4:29	13	3:22	18	1:47	
		2	5:37	7	4:41	12	3:37	17	2:11	
		1	5:49	6	4:52	11	3:50	16	2:32	

TIMETABLE A--for stars in the **SAME CELESTIAL HEMISPHERE** as the circumpolar region.

Latitude	Dec	Time	Dec	Time	Dec	Time	Dec	Time	Dec	Time
69	21	12:00	16	9:13	11	8:01	6	7:03	1	6:10
	20	10:45	15	8:57	10	7:49	5	6:52	0	6:00
	19	10:15	14	8:42	9	7:37	4	6:41		
	18	9:51	13	8:27	8	7:25	3	6:31		
	17	9:31	12	8:14	7	7:14	2	6:20		
68	22	12:00	17	9:16	12	8:06	7	7:10	2	6:19
	21	10:47	16	9:00	11	7:55	6	7:00	1	6:09
	20	10:17	15	8:46	10	7:43	5	6:50	0	6:00
	19	9:53	14	8:32	9	7:32	4	6:39		
	18	9:34	13	8:19	8	7:21	3	6:29		
67	23	12:00	18	9:19	13	8:11	8	7:17	3	6:28
	22	10:48	17	9:04	12	8:00	7	7:07	2	6:18
	21	10:18	16	8:49	11	7:49	6	6:57	1	6:09
	20	9:56	15	8:36	10	7:38	5	6:47	0	6:00
	19	9:36	14	8:23	9	7:27	4	6:37		
66	24	12:00	19	9:22	14	8:16	9	7:23	4	6:36
	23	10:49	18	9:07	13	8:04	8	7:13	3	6:27
	22	10:20	17	8:53	12	7:54	7	7:04	2	6:17
	21	9:58	16	8:40	11	7:43	6	6:54	1	6:08
	20	9:39	15	8:28	10	7:33	5	6:45	0	6:00
65	25	12:00	19	9:10	13	7:58	7	7:01	1	6:08
	24	10:50	18	8:56	12	7:48	6	6:52	0	6:00
	23	10:22	17	8:43	11	7:38	5	6:43		
	22	10:00	16	8:31	10	7:28	4	6:34		
	21	9:41	15	8:20	9	7:19	3	6:25		
	20	9:25	14	8:09	8	7:10	2	6:17		
64	26	12:00	20	9:13	14	8:02	8	7:06	2	6:16
	25	10:51	19	8:59	13	7:53	7	6:58	1	6:08
	24	10:23	18	8:47	12	7:43	6	6:49	0	6:00
	23	10:01	17	8:35	11	7:33	5	6:41		
	22	9:43	16	8:24	10	7:24	4	6:32		
	21	9:27	15	8:13	9	7:15	3	6:24		
63	27	12:00	21	9:15	15	8:06	9	7:12	3	6:23
	26	10:52	20	9:02	14	7:57	8	7:04	2	6:15
	25	10:24	19	8:50	13	7:47	7	6:55	1	6:07
	24	10:03	18	8:38	12	7:38	6	6:47	0	6:00
	23	9:45	17	8:27	11	7:29	5	6:39		
	22	9:29	16	8:16	10	7:20	4	6:31		
62	28	12:00	22	9:17	16	8:10	10	7:17	4	6:30
	27	10:53	21	9:04	15	8:01	9	7:09	3	6:22
	26	10:26	20	8:52	14	7:51	8	7:01	2	6:15
	25	10:05	19	8:41	13	7:42	7	6:53	1	6:07
	24	9:47	18	8:30	12	7:34	6	6:45	0	6:00
	23	9:31	17	8:20	11	7:25	5	6:37		

TIMETABLE B--for stars in the celestial hemisphere OPPOSITE THE CIRCUMPOLAR REGION.

Dec	Time	Dec	Time	Dec	Time	Dec	Time	Dec	Time	Latitude
1	5:49	6	4:56	11	3:58	16	2:46	21	0:00	69
0	6:00	5	5:07	10	4:10	15	3:02	20	1:14	
		4	5:18	9	4:22	14	3:17	19	1:44	
		3	5:28	8	4:34	13	3:32	18	2:08	
		2	5:39	7	4:45	12	3:45	17	2:28	
2	5:40	7	4:49	12	3:53	17	2:43	22	0:00	68
1	5:50	6	4:59	11	4:04	16	2:59	21	1:12	
0	6:00	5	5:09	10	4:16	15	3:13	20	1:42	
		4	5:20	9	4:27	14	3:27	19	2:06	
		3	5:30	8	4:38	13	3:40	18	2:25	
3	5:31	8	4:42	13	3:48	18	2:40	23	0:00	67
2	5:41	7	4:52	12	3:59	17	2:55	22	1:11	
1	5:50	6	5:02	11	4:10	16	3:10	21	1:41	
0	6:00	5	5:12	10	4:21	15	3:23	20	2:03	
		4	5:22	9	4:32	14	3:36	19	2:23	
4	5:23	9	4:36	14	3:43	19	2:37	24	0:00	66
3	5:32	8	4:46	13	3:55	18	2:52	23	1:10	
2	5:42	7	4:55	12	4:05	17	3:06	22	1:39	
1	5:51	6	5:05	11	4:16	16	3:19	21	2:01	
0	6:00	5	5:14	10	4:26	15	3:31	20	2:20	
1	5:51	7	4:58	13	4:01	19	2:49	25	0:00	65
0	6:00	6	5:07	12	4:11	18	3:03	24	1:09	
		5	5:16	11	4:21	17	3:16	23	1:37	
		4	5:25	10	4:31	16	3:28	22	1:59	
		3	5:34	9	4:40	15	3:39	21	2:18	
		2	5:42	8	4:49	14	3:50	20	2:34	
2	5:43	8	4:53	14	3:57	20	2:46	26	0:00	64
1	5:51	7	5:01	13	4:06	19	3:00	25	1:08	
0	6:00	6	5:10	12	4:16	18	3:12	24	1:36	
		5	5:18	11	4:26	17	3:24	23	1:58	
		4	5:27	10	4:35	16	3:35	22	2:16	
		3	5:35	9	4:44	15	3:46	21	2:32	
3	5:36	9	4:47	15	3:53	21	2:44	27	0:00	63
2	5:44	8	4:55	14	4:02	20	2:57	26	1:07	
1	5:52	7	5:04	13	4:12	19	3:09	25	1:35	
0	6:00	6	5:12	12	4:21	18	3:21	24	1:56	
		5	5:20	11	4:30	17	3:32	23	2:14	
		4	5:28	10	4:39	16	3:43	22	2:30	
4	5:29	10	4:42	16	3:49	22	2:42	28	0:00	62
3	5:37	9	4:50	15	3:58	21	2:55	27	1:06	
2	5:44	8	4:58	14	4:08	20	3:07	26	1:33	
1	5:52	7	5:06	13	4:17	19	3:18	25	1:54	
0	6:00	6	5:14	12	4:25	18	3:29	24	2:12	
		5	5:22	11	4:34	17	3:39	23	2:28	

TIMETABLE A--for stars in the **SAME CELESTIAL HEMISPHERE** as the circumpolar region.

Latitude	Dec	Time	Dec	Time	Dec	Time	Dec	Time	Dec	Time
61	29	12:00	23	9:19	17	8:13	11	7:22	5	6:36
	28	10:54	22	9:07	16	8:04	10	7:14	4	6:28
	27	10:27	21	8:55	15	7:55	9	7:06	3	6:21
	26	10:06	20	8:44	14	7:46	8	6:58	2	6:14
	25	9:49	19	8:33	13	7:38	7	6:51	1	6:07
	24	9:33	18	8:23	12	7:30	6	6:43	0	6:00
60	30	12:00	23	9:09	16	7:59	9	7:03	2	6:13
	29	10:55	22	8:57	15	7:50	8	6:56	1	6:06
	28	10:28	21	8:46	14	7:42	7	6:49	0	6:00
	27	10:07	20	8:36	13	7:34	6	6:41		
	26	9:50	19	8:26	12	7:26	5	6:34		
	25	9:35	18	8:16	11	7:18	4	6:27		
	24	9:21	17	8:07	10	7:11	3	6:20		
59	31	12:00	24	9:11	17	8:02	10	7:08	3	6:20
	30	10:55	23	8:59	16	7:54	9	7:01	2	6:13
	29	10:29	22	8:49	15	7:45	8	6:54	1	6:06
	28	10:08	21	8:38	14	7:38	7	6:47	0	6:00
	27	9:51	20	8:29	13	7:30	6	6:40		
	26	9:37	19	8:19	12	7:22	5	6:33		
	25	9:23	18	8:10	11	7:15	4	6:26		
58	32	12:00	25	9:13	18	8:05	11	7:12	4	6:25
	31	10:56	24	9:01	17	7:57	10	7:05	3	6:19
	30	10:30	23	8:51	16	7:49	9	6:58	2	6:12
	29	10:10	22	8:41	15	7:41	8	6:51	1	6:06
	28	9:53	21	8:31	14	7:34	7	6:45	0	6:00
	27	9:38	20	8:22	13	7:26	6	6:38		
	26	9:25	19	8:13	12	7:19	5	6:32		
57	33	12:00	26	9:14	19	8:08	12	7:16	5	6:30
	32	10:56	25	9:03	18	8:00	11	7:09	4	6:24
	31	10:30	24	8:53	17	7:52	10	7:03	3	6:18
	30	10:11	23	8:43	16	7:44	9	6:56	2	6:12
	29	9:54	22	8:33	15	7:37	8	6:49	1	6:06
	28	9:39	21	8:24	14	7:30	7	6:43	0	6:00
	27	9:26	20	8:16	13	7:23	6	6:37		
56	34	12:00	27	9:16	20	8:10	13	7:20	6	6:35
	33	10:57	26	9:05	19	8:02	12	7:13	5	6:29
	32	10:31	25	8:54	18	7:55	11	7:06	4	6:23
	31	10:11	24	8:45	17	7:47	10	7:00	3	6:17
	30	9:55	23	8:35	16	7:40	9	6:54	2	6:11
	29	9:41	22	8:27	15	7:33	8	6:48	1	6:05
	28	9:28	21	8:18	14	7:26	7	6:41	0	6:00

TIMETABLE B--for stars in the celestial hemisphere OPPOSITE THE CIRCUMPOLAR REGION.

Dec	Time	Dec	Time	Dec	Time	Dec	Time	Dec	Time	Latitude
5	5:23	11	4:37	17	3:46	23	2:40	29	0:00	61
4	5:31	10	4:45	16	3:55	22	2:52	28	1:05	
3	5:38	9	4:53	15	4:04	21	3:04	27	1:32	
2	5:45	8	5:01	14	4:13	20	3:15	26	1:53	
1	5:52	7	5:08	13	4:21	19	3:26	25	2:10	
0	6:00	6	5:16	12	4:29	18	3:36	24	2:26	
2	5:46	9	4:56	16	4:00	23	2:50	30	0:00	60
1	5:53	8	5:03	15	4:09	22	3:02	29	1:04	
0	6:00	7	5:10	14	4:17	21	3:13	28	1:31	
		6	5:18	13	4:25	20	3:23	27	1:52	
		5	5:25	12	4:33	19	3:33	26	2:09	
		4	5:32	11	4:41	18	3:43	25	2:24	
		3	5:39	10	4:48	17	3:52	24	2:38	
3	5:39	10	4:51	17	3:57	24	2:48	31	0:00	59
2	5:46	9	4:58	16	4:05	23	3:00	30	1:04	
1	5:53	8	5:05	15	4:14	22	3:10	29	1:30	
0	6:00	7	5:12	14	4:21	21	3:21	28	1:51	
		6	5:19	13	4:29	20	3:30	27	2:08	
		5	5:26	12	4:37	19	3:40	26	2:22	
		4	5:33	11	4:44	18	3:49	25	2:36	
4	5:34	11	4:47	18	3:54	25	2:46	32	0:00	58
3	5:40	10	4:54	17	4:02	24	2:58	31	1:03	
2	5:47	9	5:01	16	4:10	23	3:08	30	1:29	
1	5:53	8	5:08	15	4:18	22	3:18	29	1:49	
0	6:00	7	5:14	14	4:25	21	3:28	28	2:06	
		6	5:21	13	4:33	20	3:37	27	2:21	
		5	5:27	12	4:40	19	3:46	26	2:34	
5	5:29	12	4:43	19	3:51	26	2:45	33	0:00	57
4	5:35	11	4:50	18	3:59	25	2:56	32	1:03	
3	5:41	10	4:56	17	4:07	24	3:06	31	1:29	
2	5:47	9	5:03	16	4:15	23	3:16	30	1:48	
1	5:53	8	5:10	15	4:22	22	3:26	29	2:05	
0	6:00	7	5:16	14	4:29	21	3:35	28	2:20	
		6	5:22	13	4:36	20	3:43	27	2:33	
6	5:24	13	4:39	20	3:49	27	2:43	34	0:00	56
5	5:30	12	4:46	19	3:57	26	2:54	33	1:02	
4	5:36	11	4:53	18	4:04	25	3:05	32	1:28	
3	5:42	10	4:59	17	4:12	24	3:14	31	1:48	
2	5:48	9	5:05	16	4:19	23	3:24	30	2:04	
1	5:54	8	5:11	15	4:26	22	3:32	29	2:18	
0	6:00	7	5:18	14	4:33	21	3:41	28	2:31	

TIMETABLE A--for stars in the **SAME CELESTIAL HEMISPHERE** as the circumpolar region.

Latitude	Dec	Time	Dec	Time	Dec	Time	Dec	Time	Dec	Time
55	35	12:00	27	9:06	19	7:57	11	7:04	3	6:17
	34	10:57	26	8:56	18	7:50	10	6:58	2	6:11
	33	10:32	25	8:47	17	7:43	9	6:52	1	6:05
	32	10:12	24	8:37	16	7:36	8	6:46	0	6:00
	31	9:56	23	8:29	15	7:29	7	6:40		
	30	9:42	22	8:20	14	7:23	6	6:34		
	29	9:29	21	8:12	13	7:17	5	6:28		
	28	9:17	20	8:05	12	7:10	4	6:22		
54	36	12:00	28	9:08	20	8:00	12	7:08	4	6:22
	35	10:58	27	8:58	19	7:53	11	7:02	3	6:16
	34	10:32	26	8:48	18	7:46	10	6:56	2	6:11
	33	10:13	25	8:39	17	7:39	9	6:50	1	6:05
	32	9:57	24	8:31	16	7:32	8	6:44	0	6:00
	31	9:43	23	8:22	15	7:26	7	6:38		
	30	9:30	22	8:15	14	7:20	6	6:33		
	29	9:18	21	8:07	13	7:14	5	6:27		
53	37	12:00	29	9:09	21	8:02	13	7:11	5	6:26
	36	10:58	28	8:59	20	7:55	12	7:05	4	6:21
	35	10:33	27	8:50	19	7:48	11	6:59	3	6:15
	34	10:14	26	8:41	18	7:42	10	6:54	2	6:10
	33	9:58	25	8:32	17	7:35	9	6:48	1	6:05
	32	9:44	24	8:24	16	7:29	8	6:42	0	6:00
	31	9:31	23	8:17	15	7:23	7	6:37		
	30	9:20	22	8:09	14	7:17	6	6:32		
52	38	12:00	30	9:10	22	8:04	14	7:14	6	6:30
	37	10:58	29	9:00	21	7:57	13	7:08	5	6:25
	36	10:33	28	8:51	20	7:51	12	7:03	4	6:20
	35	10:14	27	8:42	19	7:44	11	6:57	3	6:15
	34	9:58	26	8:34	18	7:38	10	6:52	2	6:10
	33	9:44	25	8:26	17	7:32	9	6:46	1	6:05
	32	9:32	24	8:18	16	7:26	8	6:41	0	6:00
	31	9:21	23	8:11	15	7:20	7	6:36		
51	39	12:00	31	9:11	23	8:06	15	7:17	7	6:34
	38	10:59	30	9:01	22	7:59	14	7:11	6	6:29
	37	10:34	29	8:52	21	7:53	13	7:06	5	6:24
	36	10:15	28	8:44	20	7:46	12	7:00	4	6:19
	35	9:59	27	8:35	19	7:40	11	6:55	3	6:14
	34	9:45	26	8:28	18	7:34	10	6:50	2	6:09
	33	9:33	25	8:20	17	7:28	9	6:45	1	6:04
	32	9:22	24	8:13	16	7:22	8	6:39	0	6:00
50	40	12:00	31	9:02	22	7:55	13	7:03	4	6:19
	39	10:59	30	8:53	21	7:48	12	6:58	3	6:14
	38	10:34	29	8:45	20	7:42	11	6:53	2	6:09
	37	10:15	28	8:37	19	7:36	10	6:48	1	6:04
	36	9:59	27	8:29	18	7:31	9	6:43	0	6:00
	35	9:46	26	8:22	17	7:25	8	6:38		
	34	9:33	25	8:15	16	7:19	7	6:33		
	33	9:22	24	8:08	15	7:14	6	6:28		
	32	9:12	23	8:01	14	7:09	5	6:23		

TIMETABLE B--for stars in the celestial hemisphere OPPOSITE THE CIRCUMPOLAR REGION.

Dec	Time	Dec	Time	Dec	Time	Dec	Time	Dec	Time	Latitude
3	5:42	11	4:55	19	4:02	27	2:53	35	0:00	55
2	5:48	10	5:01	18	4:09	26	3:03	34	1:02	
1	5:54	9	5:07	17	4:16	25	3:12	33	1:27	
0	6:00	8	5:13	16	4:23	24	3:22	32	1:47	
		7	5:19	15	4:30	23	3:30	31	2:03	
		6	5:25	14	4:36	22	3:39	30	2:17	
		5	5:31	13	4:42	21	3:47	29	2:30	
		4	5:37	12	4:49	20	3:54	28	2:42	
4	5:37	12	4:51	20	3:59	28	2:51	36	0:00	54
3	5:43	11	4:47	19	4:06	27	3:01	35	1:01	
2	5:48	10	5:03	18	4:13	26	3:11	34	1:27	
1	5:54	9	5:09	17	4:20	25	3:20	33	1:46	
0	6:00	8	5:15	16	4:27	24	3:28	32	2:02	
		7	5:21	15	4:33	23	3:37	31	2:16	
		6	5:26	14	4:39	22	3:44	30	2:29	
		5	5:32	13	4:45	21	3:52	29	2:41	
5	5:33	13	4:48	21	3:57	29	2:50	37	0:00	53
4	5:38	12	4:54	20	4:04	28	3:00	36	1:01	
3	5:44	11	5:00	19	4:11	27	3:09	35	1:26	
2	5:49	10	5:05	18	4:17	26	3:18	34	1:45	
1	5:54	9	5:11	17	4:24	25	3:27	33	2:01	
0	6:00	8	5:17	16	4:30	24	3:35	32	2:15	
		7	5:22	15	4:36	23	3:42	31	2:28	
		6	5:27	14	4:42	22	3:50	30	2:39	
6	5:29	14	4:45	22	3:55	30	2:49	38	0:00	52
5	5:34	13	4:51	21	4:02	29	2:59	37	1:01	
4	5:39	12	4:56	20	4:08	28	3:08	36	1:26	
3	5:44	11	5:02	19	4:15	27	3:17	35	1:45	
2	5:49	10	5:07	18	4:21	26	3:25	34	2:01	
1	5:54	9	5:13	17	4:27	25	3:33	33	2:15	
0	6:00	8	5:18	16	4:33	24	3:41	32	2:27	
		7	5:23	15	4:39	23	3:48	31	2:38	
7	5:25	15	4:42	23	3:53	31	2:48	39	0:00	51
6	5:30	14	4:48	22	4:00	30	2:58	38	1:00	
5	5:35	13	4:53	21	4:06	29	3:07	37	1:25	
4	5:40	12	4:59	20	4:13	28	3:15	36	1:44	
3	5:45	11	5:04	19	4:19	27	3:24	35	2:00	
2	5:50	10	5:09	18	4:25	26	3:31	34	2:14	
1	5:55	9	5:14	17	4:31	25	3:39	33	2:26	
0	6:00	8	5:20	16	4:37	24	3:46	32	2:37	
4	5:40	13	4:56	22	4:04	31	2:57	40	0:00	50
3	5:45	12	5:01	21	4:11	30	3:06	39	1:00	
2	5:50	11	5:06	20	4:17	29	3:14	38	1:25	
1	5:55	10	5:11	19	4:23	28	3:22	37	1:44	
0	6:00	9	5:16	18	4:28	27	3:30	36	2:00	
		8	5:21	17	4:34	26	3:37	35	2:13	
		7	5:26	16	4:40	25	3:44	34	2:26	
		6	5:31	15	4:45	24	3:51	33	2:37	
		5	5:36	14	4:50	23	3:58	32	2:47	

TIMETABLE A--for stars in the **SAME CELESTIAL HEMISPHERE** as the circumpolar region.

Latitude	Dec	Time	Dec	Time	Dec	Time	Dec	Time	Dec	Time
49	41	12:00	32	9:03	23	7:56	14	7:06	5	6:23
	40	10:59	31	8:54	22	7:50	13	7:01	4	6:18
	39	10:34	30	8:46	21	7:44	12	6:56	3	6:13
	38	10:15	29	8:38	20	7:39	11	6:51	2	6:09
	37	10:00	28	8:30	19	7:33	10	6:46	1	6:04
	36	9:46	27	8:23	18	7:27	9	6:41	0	6:00
	35	9:34	26	8:16	17	7:22	8	6:37		
	34	9:23	25	8:09	16	7:17	7	6:32		
	33	9:13	24	8:03	15	7:11	6	6:27		
48	42	12:00	33	9:04	24	7:58	15	7:09	6	6:26
	41	10:59	32	8:55	23	7:52	14	7:04	5	6:22
	40	10:34	31	8:47	22	7:46	13	6:59	4	6:17
	39	10:16	30	8:39	21	7:40	12	6:54	3	6:13
	38	10:00	29	8:31	20	7:35	11	6:49	2	6:08
	37	9:47	28	8:24	19	7:29	10	6:45	1	6:04
	36	9:35	27	8:17	18	7:24	9	6:40	0	6:00
	35	9:24	26	8:11	17	7:19	8	6:35		
	34	9:14	25	8:04	16	7:14	7	6:31		
47	43	12:00	34	9:05	25	8:00	16	7:11	7	6:30
	42	10:59	33	8:56	24	7:54	15	7:06	6	6:25
	41	10:35	32	8:48	23	7:48	14	7:02	5	6:21
	40	10:16	31	8:40	22	7:42	13	6:57	4	6:17
	39	10:01	30	8:33	21	7:37	12	6:52	3	6:12
	38	9:47	29	8:25	20	7:31	11	6:48	2	6:08
	37	9:35	28	8:19	19	7:26	10	6:43	1	6:04
	36	9:24	27	8:12	18	7:21	9	6:39	0	6:00
	35	9:14	26	8:06	17	7:16	8	6:34		
46	44	12:00	35	9:05	26	8:01	17	7:13	8	6:33
	43	10:59	34	8:57	25	7:55	16	7:09	7	6:29
	42	10:35	33	8:49	24	7:49	15	7:04	6	6:24
	41	10:16	32	8:41	23	7:44	14	6:59	5	6:20
	40	10:01	31	8:33	22	7:38	13	6:55	4	6:16
	39	9:47	30	8:26	21	7:33	12	6:50	3	6:12
	38	9:36	29	8:20	20	7:28	11	6:46	2	6:08
	37	9:25	28	8:13	19	7:23	10	6:42	1	6:04
	36	9:15	27	8:07	18	7:18	9	6:37	0	6:00
45	45	12:00	35	8:57	25	7:51	15	7:02	5	6:20
	44	10:59	34	8:49	24	7:45	14	6:57	4	6:16
	43	10:35	33	8:41	23	7:40	13	6:53	3	6:12
	42	10:16	32	8:34	22	7:35	12	6:49	2	6:08
	41	10:01	31	8:27	21	7:30	11	6:44	1	6:04
	40	9:48	30	8:21	20	7:25	10	6:40	0	6:00
	39	9:36	29	8:14	19	7:20	9	6:36		
	38	9:25	28	8:08	18	7:15	8	6:32		
	37	9:15	27	8:02	17	7:11	7	6:28		
	36	9:06	26	7:56	16	7:06	6	6:24		

TIMETABLE B--for stars in the celestial hemisphere OPPOSITE THE CIRCUMPOLAR REGION.

Dec	Time	Dec	Time	Dec	Time	Dec	Time	Dec	Time	Latitude
5	5:36	14	4:53	23	4:03	32	2:56	41	0:00	49
4	5:41	13	4:58	22	4:09	31	3:05	40	1:00	
3	5:46	12	5:03	21	4:15	30	3:13	39	1:25	
2	5:50	11	5:08	20	4:20	29	3:21	38	1:44	
1	5:55	10	5:13	19	4:26	28	3:29	37	1:59	
0	6:00	9	5:18	18	4:32	27	3:36	36	2:13	
		8	5:22	17	4:37	26	3:43	35	2:25	
		7	5:27	16	4:42	25	3:50	34	2:36	
		6	5:32	15	4:48	24	3:56	33	2:46	
6	5:33	15	4:50	24	4:01	33	2:55	42	0:00	48
5	5:37	14	4:55	23	4:07	32	3:04	41	1:00	
4	5:42	13	5:00	22	4:13	31	3:12	40	1:25	
3	5:46	12	5:05	21	4:19	30	3:20	39	1:43	
2	5:51	11	5:10	20	4:24	29	3:28	38	1:59	
1	5:55	10	5:14	19	4:30	28	3:35	37	2:12	
0	6:00	9	5:19	18	4:35	27	3:42	36	2:24	
		8	5:24	17	4:40	26	3:48	35	2:35	
		7	5:28	16	4:45	25	3:55	34	2:45	
7	5:29	16	4:48	25	3:59	34	2:54	43	0:00	47
6	5:34	15	4:53	24	4:05	33	3:03	42	1:00	
5	5:38	14	4:57	23	4:11	32	3:11	41	1:24	
4	5:42	13	5:02	22	4:17	31	3:19	40	1:43	
3	5:47	12	5:07	21	4:22	30	3:26	39	1:58	
2	5:51	11	5:11	20	4:28	29	3:34	38	2:12	
1	5:55	10	5:16	19	4:33	28	3:40	37	2:24	
0	6:00	9	5:20	18	4:38	27	3:47	36	2:35	
		8	5:25	17	4:43	26	3:53	35	2:45	
8	5:26	17	4:46	26	3:58	35	2:54	44	0:00	46
7	5:30	16	4:50	25	4:04	34	3:02	43	1:00	
6	5:35	15	4:55	24	4:10	33	3:10	42	1:24	
5	5:39	14	5:00	23	4:15	32	3:18	41	1:43	
4	5:43	13	5:04	22	4:21	31	3:26	40	1:58	
3	5:47	12	5:09	21	4:26	30	3:33	39	2:12	
2	5:51	11	5:13	20	4:31	29	3:39	38	2:23	
1	5:55	10	5:17	19	4:36	28	3:46	37	2:34	
0	6:00	9	5:22	18	4:41	27	3:52	36	2:44	
5	5:39	15	4:57	25	4:08	35	3:02	45	0:00	45
4	5:43	14	5:02	24	4:14	34	3:10	44	1:00	
3	5:47	13	5:06	23	4:19	33	3:18	43	1:24	
2	5:51	12	5:10	22	4:24	32	3:25	42	1:43	
1	5:55	11	5:15	21	4:29	31	3:32	41	1:58	
0	6:00	10	5:19	20	4:34	30	3:38	40	2:11	
		9	5:23	19	4:39	29	3:45	39	2:23	
		8	5:27	18	4:44	28	3:51	38	2:34	
		7	5:31	17	4:48	27	3:57	37	2:44	
		6	5:35	16	4:53	26	4:03	36	2:53	

TIMETABLE A--for stars in the **SAME CELESTIAL HEMISPHERE** as the circumpolar region.

Latitude	Dec	Time	Dec	Time	Dec	Time	Dec	Time	Dec	Time
44	46	12:00	36	8:58	26	7:52	16	7:04	6	6:23
	45	10:59	35	8:50	25	7:47	15	6:59	5	6:19
	44	10:35	34	8:42	24	7:41	14	6:55	4	6:15
	43	10:16	33	8:35	23	7:36	13	6:51	3	6:11
	42	10:01	32	8:28	22	7:31	12	6:47	2	6:07
	41	9:48	31	8:21	21	7:27	11	6:43	1	6:03
	40	9:36	30	8:15	20	7:22	10	6:39	0	6:00
	39	9:25	29	8:09	19	7:17	9	6:35		
	38	9:15	28	8:03	18	7:13	8	6:31		
	37	9:06	27	7:57	17	7:08	7	6:27		
43	47	12:00	37	8:58	27	7:53	17	7:06	7	6:26
	46	10:59	36	8:50	26	7:48	16	7:02	6	6:22
	45	10:35	35	8:43	25	7:43	15	6:57	5	6:18
	44	10:16	34	8:35	24	7:38	14	6:53	4	6:14
	43	10:01	33	8:29	23	7:33	13	6:49	3	6:11
	42	9:48	32	8:22	22	7:28	12	6:45	2	6:07
	41	9:36	31	8:16	21	7:23	11	6:41	1	6:03
	40	9:25	30	8:10	20	7:19	10	6:37	0	6:00
	39	9:16	29	8:04	19	7:14	9	6:33		
	38	9:07	28	7:58	18	7:10	8	6:30		
42	48	12:00	38	8:58	28	7:54	18	7:08	8	6:29
	47	10:59	37	8:50	27	7:49	17	7:03	7	6:25
	46	10:35	36	8:43	26	7:44	16	6:59	6	6:21
	45	10:16	35	8:36	25	7:39	15	6:55	5	6:18
	44	10:01	34	8:29	24	7:34	14	6:51	4	6:14
	43	9:48	33	8:23	23	7:29	13	6:47	3	6:10
	42	9:36	32	8:16	22	7:25	12	6:44	2	6:07
	41	9:26	31	8:11	21	7:20	11	6:40	1	6:03
	40	9:16	30	8:05	20	7:16	10	6:36	0	6:00
	39	9:07	29	7:59	19	7:12	9	6:32		
41	49	12:00	39	8:58	29	7:55	19	7:09	9	6:31
	48	10:59	38	8:51	28	7:50	18	7:05	8	6:28
	47	10:35	37	8:43	27	7:45	17	7:01	7	6:24
	46	10:16	36	8:36	26	7:40	16	6:57	6	6:20
	45	10:01	35	8:29	25	7:35	15	6:53	5	6:17
	44	9:48	34	8:23	24	7:31	14	6:50	4	6:13
	43	9:36	33	8:17	23	7:26	13	6:46	3	6:10
	42	9:26	32	8:11	22	7:22	12	6:42	2	6:06
	41	9:16	31	8:05	21	7:17	11	6:38	1	6:03
	40	9:07	30	8:00	20	7:13	10	6:35	0	6:00

TIMETABLE B--for stars in the celestial hemisphere OPPOSITE THE CIRCUMPOLAR REGION.

Dec	Time	Dec	Time	Dec	Time	Dec	Time	Dec	Time	Latitude
6	5:36	16	4:55	26	4:07	36	3:01	46	0:00	44
5	5:40	15	5:00	25	4:12	35	3:09	45	1:00	
4	5:44	14	5:04	24	4:18	34	3:17	44	1:24	
3	5:48	13	5:08	23	4:23	33	3:24	43	1:43	
2	5:52	12	5:12	22	4:28	32	3:31	42	1:58	
1	5:56	11	5:16	21	4:32	31	3:38	41	2:11	
0	6:00	10	5:20	20	4:37	30	3:44	40	2:23	
		9	5:24	19	4:42	29	3:50	39	2:34	
		8	5:28	18	4:46	28	3:56	38	2:44	
		7	5:32	17	4:51	27	4:02	37	2:53	
7	5:33	17	4:53	27	4:06	37	3:01	47	0:00	43
6	5:37	16	4:57	26	4:11	36	3:09	46	1:00	
5	5:41	15	5:02	25	4:16	35	3:16	45	1:24	
4	5:45	14	5:06	24	4:21	34	3:24	44	1:43	
3	5:48	13	5:10	23	4:26	33	3:30	43	1:58	
2	5:52	12	5:14	22	4:31	32	3:37	42	2:11	
1	5:56	11	5:18	21	4:36	31	3:43	41	2:23	
0	6:00	10	5:22	20	4:40	30	3:49	40	2:34	
		9	5:26	19	4:45	29	3:55	39	2:43	
		8	5:29	18	4:49	28	4:01	38	2:52	
8	5:30	18	4:51	28	4:05	38	3:01	48	0:00	42
7	5:34	17	4:56	27	4:10	37	3:09	47	1:00	
6	5:38	16	5:00	26	4:15	36	3:16	46	1:24	
5	5:41	15	5:04	25	4:20	35	3:23	45	1:43	
4	5:45	14	5:08	24	4:25	34	3:30	44	1:58	
3	5:49	13	5:12	23	4:30	33	3:36	43	2:11	
2	5:52	12	5:15	22	4:34	32	3:43	42	2:23	
1	5:56	11	5:19	21	4:39	31	3:48	41	2:33	
0	6:00	10	5:23	20	4:43	30	3:54	40	2:43	
		9	5:27	19	4:47	29	4:00	39	2:52	
9	5:28	19	4:50	29	4:04	39	3:01	49	0:00	41
8	5:31	18	4:54	28	4:09	38	3:08	48	1:00	
7	5:35	17	4:58	27	4:14	37	3:16	47	1:24	
6	5:39	16	5:02	26	4:19	36	3:23	46	1:43	
5	5:42	15	5:06	25	4:24	35	3:30	45	1:58	
4	5:46	14	5:09	24	4:28	34	3:36	44	2:11	
3	5:49	13	5:13	23	4:33	33	3:42	43	2:23	
2	5:53	12	5:17	22	4:37	32	3:48	42	2:33	
1	5:56	11	5:21	21	4:42	31	3:54	41	2:43	
0	6:00	10	5:24	20	4:46	30	3:59	40	2:52	

TIMETABLE A--for stars in the **SAME CELESTIAL HEMISPHERE** as the circumpolar region.

Latitude	Dec	Time	Dec	Time	Dec	Time	Dec	Time	Dec	Time
40	50	12:00	39	8:51	28	7:45	17	6:59	6	6:20
	49	10:59	38	8:43	27	7:41	16	6:55	5	6:16
	48	10:34	37	8:36	26	7:36	15	6:51	4	6:13
	47	10:16	36	8:30	25	7:32	14	6:48	3	6:10
	46	10:01	35	8:23	24	7:27	13	6:44	2	6:06
	45	9:48	34	8:17	23	7:23	12	6:41	1	6:03
	44	9:36	33	8:12	22	7:19	11	6:37	0	6:00
	43	9:25	32	8:06	21	7:15	10	6:34		
	42	9:16	31	8:01	20	7:11	9	6:30		
	41	9:07	30	7:55	19	7:07	8	6:27		
	40	8:59	29	7:50	18	7:03	7	6:23		
39	51	12:00	40	8:51	29	7:46	18	7:01	7	6:22
	50	10:59	39	8:43	28	7:42	17	6:57	6	6:19
	49	10:34	38	8:36	27	7:37	16	6:53	5	6:16
	48	10:16	37	8:30	26	7:33	15	6:50	4	6:12
	47	10:01	36	8:24	25	7:28	14	6:46	3	6:09
	46	9:47	35	8:18	24	7:24	13	6:43	2	6:06
	45	9:36	34	8:12	23	7:20	12	6:39	1	6:03
	44	9:25	33	8:06	22	7:16	11	6:36	0	6:00
	43	9:16	32	8:01	21	7:12	10	6:32		
	42	9:07	31	7:56	20	7:08	9	6:29		
	41	8:58	30	7:51	19	7:04	8	6:26		
38	52	12:00	41	8:51	30	7:47	19	7:02	8	6:25
	51	10:59	40	8:43	29	7:42	18	6:58	7	6:22
	50	10:34	39	8:36	28	7:38	17	6:55	6	6:18
	49	10:15	38	8:30	27	7:33	16	6:51	5	6:15
	48	10:00	37	8:24	26	7:29	15	6:48	4	6:12
	47	9:47	36	8:18	25	7:25	14	6:44	3	6:09
	46	9:36	35	8:12	24	7:21	13	6:41	2	6:06
	45	9:25	34	8:07	23	7:17	12	6:38	1	6:03
	44	9:15	33	8:01	22	7:13	11	6:34	0	6:00
	43	9:07	32	7:56	21	7:09	10	6:31		
	42	8:58	31	7:51	20	7:06	9	6:28		
37	53	12:00	42	8:50	31	7:47	20	7:03	9	6:27
	52	10:58	41	8:43	30	7:43	19	7:00	8	6:24
	51	10:34	40	8:36	29	7:38	18	6:56	7	6:21
	50	10:15	39	8:30	28	7:34	17	6:53	6	6:18
	49	10:00	38	8:24	27	7:30	16	6:49	5	6:15
	48	9:47	37	8:18	26	7:26	15	6:46	4	6:12
	47	9:35	36	8:12	25	7:22	14	6:43	3	6:09
	46	9:25	35	8:07	24	7:18	13	6:40	2	6:06
	45	9:15	34	8:02	23	7:14	12	6:36	1	6:03
	44	9:06	33	7:57	22	7:10	11	6:33	0	6:00
	43	8:58	32	7:52	21	7:07	10	6:30		

TIMETABLE B--for stars in the celestial hemisphere OPPOSITE THE CIRCUMPOLAR REGION.

Dec	Time	Dec	Time	Dec	Time	Dec	Time	Dec	Time	Latitude
6	5:39	17	5:00	28	4:14	39	3:08	50	0:00	40
5	5:43	16	5:04	27	4:18	38	3:16	49	1:00	
4	5:46	15	5:08	26	4:23	37	3:23	48	1:25	
3	5:49	14	5:11	25	4:27	36	3:29	47	1:43	
2	5:53	13	5:15	24	4:32	35	3:36	46	1:58	
1	5:56	12	5:18	23	4:36	34	3:42	45	2:11	
0	6:00	11	5:22	22	4:40	33	3:47	44	2:23	
		10	5:25	21	4:44	32	3:53	43	2:34	
		9	5:29	20	4:48	31	3:58	42	2:43	
		8	5:32	19	4:52	30	4:04	41	2:52	
		7	5:36	18	4:56	29	4:09	40	3:00	
7	5:37	18	4:58	29	4:13	40	3:08	51	0:00	39
6	5:40	17	5:02	28	4:17	39	3:16	50	1:00	
5	5:43	16	5:06	27	4:22	38	3:23	49	1:25	
4	5:47	15	5:09	26	4:26	37	3:29	48	1:43	
3	5:50	14	5:13	25	4:31	36	3:35	47	1:58	
2	5:53	13	5:16	24	4:35	35	3:41	46	2:12	
1	5:56	12	5:20	23	4:39	34	3:47	45	2:23	
0	6:00	11	5:23	22	4:43	33	3:53	44	2:34	
		10	5:27	21	4:47	32	3:58	43	2:43	
		9	5:30	20	4:51	31	4:03	42	2:52	
		8	5:33	19	4:55	30	4:08	41	3:01	
8	5:34	19	4:57	30	4:12	41	3:08	52	0:00	38
7	5:37	18	5:01	29	4:17	40	3:16	51	1:00	
6	5:41	17	5:04	28	4:21	39	3:23	50	1:25	
5	5:44	16	5:08	27	4:26	38	3:29	49	1:44	
4	5:47	15	5:11	26	4:30	37	3:35	48	1:59	
3	5:50	14	5:15	25	4:34	36	3:41	47	2:12	
2	5:53	13	5:18	24	4:38	35	3:47	46	2:23	
1	5:56	12	5:21	23	4:42	34	3:52	45	2:34	
0	6:00	11	5:25	22	4:46	33	3:58	44	2:44	
		10	5:28	21	4:50	32	4:03	43	2:52	
		9	5:31	20	4:53	31	4:08	42	3:01	
9	5:32	20	4:56	31	4:12	42	3:09	53	0:00	37
8	5:35	19	4:59	30	4:16	41	3:16	52	1:01	
7	5:38	18	5:03	29	4:21	40	3:23	51	1:25	
6	5:41	17	5:06	28	4:25	39	3:29	50	1:44	
5	5:44	16	5:10	27	4:29	38	3:35	49	1:59	
4	5:47	15	5:13	26	4:33	37	3:41	48	2:12	
3	5:50	14	5:16	25	4:37	36	3:47	47	2:24	
2	5:53	13	5:19	24	4:41	35	3:52	46	2:34	
1	5:56	12	5:23	23	4:45	34	3:57	45	2:44	
0	6:00	11	5:26	22	4:49	33	4:02	44	2:53	
		10	5:29	21	4:52	32	4:07	43	3:01	

TIMETABLE A--for stars in the **SAME CELESTIAL HEMISPHERE** as the circumpolar region.

Latitude	Dec	Time	Dec	Time	Dec	Time	Dec	Time	Dec	Time
36	54	12:00	42	8:43	30	7:39	18	6:54	6	6:17
	53	10:58	41	8:36	29	7:34	17	6:51	5	6:14
	52	10:33	40	8:30	28	7:30	16	6:48	4	6:11
	51	10:15	39	8:24	27	7:26	15	6:44	3	6:08
	50	9:59	38	8:18	26	7:23	14	6:41	2	6:05
	49	9:46	37	8:12	25	7:19	13	6:38	1	6:02
	48	9:35	36	8:07	24	7:15	12	6:35	0	6:00
	47	9:24	35	8:02	23	7:11	11	6:32		
	46	9:15	34	7:57	22	7:08	10	6:29		
	45	9:06	33	7:52	21	7:04	9	6:26		
	44	8:58	32	7:48	20	7:01	8	6:23		
	43	8:50	31	7:43	19	6:57	7	6:20		
35	55	12:00	43	8:43	31	7:39	19	6:55	7	6:19
	54	10:58	42	8:36	30	7:35	18	6:52	6	6:16
	53	10:33	41	8:29	29	7:31	17	6:49	5	6:14
	52	10:14	40	8:23	28	7:27	16	6:46	4	6:11
	51	9:59	39	8:18	27	7:23	15	6:43	3	6:08
	50	9:46	38	8:12	26	7:19	14	6:40	2	6:05
	49	9:34	37	8:07	25	7:16	13	6:37	1	6:02
	48	9:24	36	8:02	24	7:12	12	6:34	0	6:00
	47	9:14	35	7:57	23	7:09	11	6:31		
	46	9:05	34	7:52	22	7:05	10	6:28		
	45	8:57	33	7:48	21	7:02	9	6:25		
	44	8:50	32	7:43	20	6:59	8	6:22		
34	56	12:00	44	8:42	32	7:39	20	6:56	8	6:21
	55	10:57	43	8:35	31	7:35	19	6:53	7	6:19
	54	10:32	42	8:29	30	7:31	18	6:50	6	6:16
	53	10:14	41	8:23	29	7:27	17	6:47	5	6:13
	52	9:58	40	8:17	28	7:24	16	6:44	4	6:10
	51	9:45	39	8:12	27	7:20	15	6:41	3	6:08
	50	9:33	38	8:07	26	7:16	14	6:38	2	6:05
	49	9:23	37	8:02	25	7:13	13	6:35	1	6:02
	48	9:14	36	7:57	24	7:09	12	6:32	0	6:00
	47	9:05	35	7:52	23	7:06	11	6:30		
	46	8:57	34	7:48	22	7:03	10	6:27		
	45	8:49	33	7:43	21	7:00	9	6:24		
33	57	12:00	45	8:41	33	7:39	21	6:57	9	6:23
	56	10:57	44	8:35	32	7:35	20	6:54	8	6:20
	55	10:32	43	8:29	31	7:31	19	6:51	7	6:18
	54	10:13	42	8:23	30	7:28	18	6:48	6	6:15
	53	9:58	41	8:17	29	7:24	17	6:45	5	6:13
	52	9:44	40	8:12	28	7:20	16	6:42	4	6:10
	51	9:33	39	8:06	27	7:17	15	6:40	3	6:07
	50	9:22	38	8:01	26	7:13	14	6:37	2	6:05
	49	9:13	37	7:57	25	7:10	13	6:34	1	6:02
	48	9:04	36	7:52	24	7:07	12	6:31	0	6:00
	47	8:56	35	7:48	23	7:04	11	6:29		
	46	8:49	34	7:43	22	7:00	10	6:26		

TIMETABLE B--for stars in the celestial hemisphere OPPOSITE THE CIRCUMPOLAR REGION.

Dec	Time	Dec	Time	Dec	Time	Dec	Time	Dec	Time	Latitude
6	5:42	18	5:05	30	4:20	42	3:16	54	0:00	36
5	5:45	17	5:08	29	4:25	41	3:23	53	1:01	
4	5:48	16	5:11	28	4:29	40	3:29	52	1:26	
3	5:51	15	5:15	27	4:33	39	3:35	51	1:44	
2	5:54	14	5:18	26	4:36	38	3:41	50	2:00	
1	5:57	13	5:21	25	4:40	37	3:47	49	2:13	
0	6:00	12	5:24	24	4:44	36	3:52	48	2:24	
		11	5:27	23	4:48	35	3:57	47	2:35	
		10	5:30	22	4:51	34	4:02	46	2:44	
		9	5:33	21	4:55	33	4:07	45	2:53	
		8	5:36	20	4:58	32	4:11	44	3:01	
		7	5:39	19	5:02	31	4:16	43	3:09	
7	5:40	19	5:04	31	4:20	43	3:16	55	0:00	35
6	5:43	18	5:07	30	4:24	42	3:23	54	1:01	
5	5:45	17	5:10	29	4:28	41	3:30	53	1:26	
4	5:48	16	5:13	28	4:32	40	3:36	52	1:45	
3	5:51	15	5:16	27	4:36	39	3:41	51	2:00	
2	5:54	14	5:19	26	4:40	38	3:47	50	2:13	
1	5:57	13	5:22	25	4:43	37	3:52	49	2:25	
0	6:00	12	5:25	24	4:47	36	3:57	48	2:35	
		11	5:28	23	4:50	35	4:02	47	2:45	
		10	5:31	22	4:54	34	4:07	46	2:54	
		9	5:34	21	4:57	33	4:11	45	3:02	
		8	5:37	20	5:00	32	4:16	44	3:09	
8	5:38	20	5:03	32	4:20	44	3:17	56	0:00	34
7	5:40	19	5:06	31	4:24	43	3:24	55	1:02	
6	5:43	18	5:09	30	4:28	42	3:30	54	1:27	
5	5:46	17	5:12	29	4:32	41	3:36	53	1:45	
4	5:49	16	5:15	28	4:35	40	3:42	52	2:01	
3	5:51	15	5:18	27	4:39	39	3:47	51	2:14	
2	5:54	14	5:21	26	4:43	38	3:52	50	2:26	
1	5:57	13	5:24	25	4:46	37	3:57	49	2:36	
0	6:00	12	5:27	24	4:50	36	4:02	48	2:45	
		11	5:29	23	4:53	35	4:07	47	2:54	
		10	5:32	22	4:56	34	4:11	46	3:02	
		9	5:35	21	4:59	33	4:16	45	3:10	
9	5:36	21	5:02	33	4:20	45	3:18	57	0:00	33
8	5:39	20	5:05	32	4:24	44	3:24	56	1:02	
7	5:41	19	5:08	31	4:28	43	3:30	55	1:27	
6	5:44	18	5:11	30	4:31	42	3:36	54	1:46	
5	5:46	17	5:14	29	4:35	41	3:42	53	2:01	
4	5:49	16	5:17	28	4:39	40	3:47	52	2:15	
3	5:52	15	5:19	27	4:42	39	3:53	51	2:26	
2	5:54	14	5:22	26	4:46	38	3:58	50	2:37	
1	5:57	13	5:25	25	4:49	37	4:02	49	2:46	
0	6:00	12	5:28	24	4:52	36	4:07	48	2:55	
		11	5:30	23	4:55	35	4:11	47	3:03	
		10	5:33	22	4:59	34	4:16	46	3:10	

TIMETABLE A--for stars in the **SAME CELESTIAL HEMISPHERE** as the circumpolar region.

Latitude	Dec	Time	Dec	Time	Dec	Time	Dec	Time	Dec	Time
32	58	12:00	46	8:41	34	7:39	22	6:58	10	6:25
	57	10:56	45	8:34	33	7:35	21	6:55	9	6:22
	56	10:31	44	8:28	32	7:31	20	6:52	8	6:20
	55	10:12	43	8:22	31	7:28	19	6:49	7	6:17
	54	9:57	42	8:16	30	7:24	18	6:46	6	6:15
	53	9:44	41	8:11	29	7:21	17	6:44	5	6:12
	52	9:32	40	8:06	28	7:17	16	6:41	4	6:10
	51	9:22	39	8:01	27	7:14	15	6:38	3	6:07
	50	9:12	38	7:56	26	7:10	14	6:35	2	6:05
	49	9:03	37	7:52	25	7:07	13	6:33	1	6:02
	48	8:55	36	7:48	24	7:04	12	6:30	0	6:00
	47	8:48	35	7:43	23	7:01	11	6:27		
31	59	12:00	47	8:40	35	7:39	23	6:59	11	6:26
	58	10:56	46	8:33	34	7:35	22	6:56	10	6:24
	57	10:30	45	8:27	33	7:31	21	6:53	9	6:21
	56	10:11	44	8:21	32	7:28	20	6:50	8	6:19
	55	9:56	43	8:16	31	7:24	19	6:47	7	6:16
	54	9:43	42	8:11	30	7:21	18	6:45	6	6:14
	53	9:31	41	8:05	29	7:17	17	6:42	5	6:12
	52	9:21	40	8:01	28	7:14	16	6:39	4	6:09
	51	9:11	39	7:56	27	7:11	15	6:37	3	6:07
	50	9:02	38	7:51	26	7:08	14	6:34	2	6:04
	49	8:54	37	7:47	25	7:05	13	6:31	1	6:02
	48	8:47	36	7:43	24	7:02	12	6:29	0	6:00
30	60	12:00	47	8:33	34	7:31	21	6:51	8	6:18
	59	10:55	46	8:26	33	7:28	20	6:48	7	6:16
	58	10:30	45	8:21	32	7:24	19	6:45	6	6:13
	57	10:11	44	8:15	31	7:21	18	6:43	5	6:11
	56	9:55	43	8:10	30	7:17	17	6:40	4	6:09
	55	9:42	42	8:05	29	7:14	16	6:38	3	6:06
	54	9:30	41	8:00	28	7:11	15	6:35	2	6:04
	53	9:20	40	7:55	27	7:08	14	6:33	1	6:02
	52	9:10	39	7:51	26	7:05	13	6:30	0	6:00
	51	9:01	38	7:47	25	7:02	12	6:28		
	50	8:53	37	7:43	24	6:59	11	6:25		
	49	8:46	36	7:39	23	6:56	10	6:23		
	48	8:39	35	7:35	22	6:53	9	6:20		
29	61	12:00	48	8:31	35	7:31	22	6:51	9	6:20
	60	10:55	47	8:25	34	7:27	21	6:49	8	6:17
	59	10:29	46	8:20	33	7:24	20	6:46	7	6:15
	58	10:10	45	8:14	32	7:21	19	6:44	6	6:13
	57	9:54	44	8:09	31	7:17	18	6:41	5	6:11
	56	9:41	43	8:04	30	7:14	17	6:39	4	6:08
	55	9:29	42	7:59	29	7:11	16	6:36	3	6:06
	54	9:18	41	7:55	28	7:08	15	6:34	2	6:04
	53	9:09	40	7:50	27	7:05	14	6:31	1	6:02
	52	9:00	39	7:46	26	7:02	13	6:29	0	6:00
	51	8:52	38	7:42	25	6:59	12	6:27		
	50	8:45	37	7:38	24	6:57	11	6:24		
	49	8:38	36	7:34	23	6:54	10	6:22		

TIMETABLE B--for stars in the celestial hemisphere OPPOSITE THE CIRCUMPOLAR REGION.

Dec	Time	Dec	Time	Dec	Time	Dec	Time	Dec	Time	Latitude
10	5:34	22	5:01	34	4:20	46	3:18	58	0:00	32
9	5:37	21	5:04	33	4:24	45	3:25	57	1:03	
8	5:39	20	5:07	32	4:28	44	3:31	56	1:28	
7	5:42	19	5:10	31	4:31	43	3:37	55	1:47	
6	5:44	18	5:13	30	4:35	42	3:43	54	2:02	
5	5:47	17	5:15	29	4:38	41	3:48	53	2:15	
4	5:49	16	5:18	28	4:42	40	3:53	52	2:27	
3	5:52	15	5:21	27	4:45	39	3:58	51	2:37	
2	5:54	14	5:24	26	4:49	38	4:03	50	2:47	
1	5:57	13	5:26	25	4:52	37	4:07	49	2:56	
0	6:00	12	5:29	24	4:55	36	4:11	48	3:04	
		11	5:32	23	4:58	35	4:16	47	3:11	
11	5:33	23	5:00	35	4:20	47	3:19	59	0:00	31
10	5:35	22	5:03	34	4:24	46	3:26	58	1:03	
9	5:38	21	5:06	33	4:28	45	3:32	57	1:29	
8	5:40	20	5:09	32	4:31	44	3:38	56	1:48	
7	5:43	19	5:12	31	4:35	43	3:43	55	2:03	
6	5:45	18	5:14	30	4:38	42	3:48	54	2:16	
5	5:47	17	5:17	29	4:42	41	3:54	53	2:28	
4	5:50	16	5:20	28	4:45	40	3:58	52	2:38	
3	5:52	15	5:22	27	4:48	39	4:03	51	2:48	
2	5:55	14	5:25	26	4:51	38	4:08	50	2:57	
1	5:57	13	5:28	25	4:54	37	4:12	49	3:05	
0	6:00	12	5:30	24	4:57	36	4:16	48	3:12	
8	5:41	21	5:08	34	4:28	47	3:26	60	0:00	30
7	5:43	20	5:11	33	4:31	46	3:33	59	1:04	
6	5:46	19	5:14	32	4:35	45	3:38	58	1:29	
5	5:48	18	5:16	31	4:38	44	3:44	57	1:48	
4	5:50	17	5:19	30	4:42	43	3:49	56	2:04	
3	5:53	16	5:21	29	4:45	42	3:54	55	2:17	
2	5:55	15	5:24	28	4:48	41	3:59	54	2:29	
1	5:57	14	5:26	27	4:51	40	4:04	53	2:39	
0	6:00	13	5:29	26	4:54	39	4:08	52	2:49	
		12	5:31	25	4:57	38	4:12	51	2:58	
		11	5:34	24	5:00	37	4:16	50	3:06	
		10	5:36	23	5:03	36	4:20	49	3:13	
		9	5:39	22	5:06	35	4:24	48	3:20	
9	5:39	22	5:08	35	4:28	48	3:28	61	0:00	29
8	5:42	21	5:10	34	4:32	47	3:34	60	1:04	
7	5:44	20	5:13	33	4:35	46	3:39	59	1:30	
6	5:46	19	5:15	32	4:38	45	3:45	58	1:49	
5	5:48	18	5:18	31	4:42	44	3:50	57	2:05	
4	5:51	17	5:20	30	4:45	43	3:55	56	2:18	
3	5:53	16	5:23	29	4:48	42	4:00	55	2:30	
2	5:55	15	5:25	28	4:51	41	4:04	54	2:41	
1	5:57	14	5:28	27	4:54	40	4:09	53	2:50	
0	6:00	13	5:30	26	4:57	39	4:13	52	2:59	
		12	5:32	25	5:00	38	4:17	51	3:07	
		11	5:35	24	5:02	37	4:21	50	3:14	
		10	5:37	23	5:05	36	4:25	49	3:21	

TIMETABLE A--for stars in the **SAME CELESTIAL HEMISPHERE** as the circumpolar region.

Latitude	Dec	Time	Dec	Time	Dec	Time	Dec	Time	Dec	Time
28	62	12:00	49	8:30	36	7:30	23	6:52	10	6:21
	61	10:54	48	8:24	35	7:27	22	6:49	9	6:19
	60	10:28	47	8:19	34	7:24	21	6:47	8	6:17
	59	10:08	46	8:13	33	7:20	20	6:44	7	6:14
	58	9:53	45	8:08	32	7:17	19	6:42	6	6:12
	57	9:39	44	8:03	31	7:14	18	6:39	5	6:10
	56	9:28	43	7:58	30	7:11	17	6:37	4	6:08
	55	9:17	42	7:54	29	7:08	16	6:35	3	6:06
	54	9:08	41	7:50	28	7:05	15	6:32	2	6:04
	53	8:59	40	7:45	27	7:02	14	6:30	1	6:02
	52	8:51	39	7:42	26	7:00	13	6:28	0	6:00
	51	8:44	38	7:38	25	6:57	12	6:25		
	50	8:37	37	7:34	24	6:54	11	6:23		
27	63	12:00	50	8:29	37	7:30	24	6:52	11	6:22
	62	10:53	49	8:23	36	7:26	23	6:49	10	6:20
	61	10:27	48	8:17	35	7:23	22	6:47	9	6:18
	60	10:07	47	8:12	34	7:20	21	6:45	8	6:16
	59	9:51	46	8:07	33	7:17	20	6:42	7	6:14
	58	9:38	45	8:02	32	7:14	19	6:40	6	6:12
	57	9:26	44	7:57	31	7:11	18	6:38	5	6:10
	56	9:16	43	7:53	30	7:08	17	6:35	4	6:08
	55	9:06	42	7:49	29	7:05	16	6:33	3	6:06
	54	8:58	41	7:45	28	7:02	15	6:31	2	6:04
	53	8:50	40	7:41	27	7:00	14	6:29	1	6:02
	52	8:42	39	7:37	26	6:57	13	6:27	0	6:00
	51	8:35	38	7:33	25	6:54	12	6:24		
26	64	12:00	51	8:28	38	7:29	25	6:52	12	6:23
	63	10:52	50	8:22	37	7:26	24	6:50	11	6:21
	62	10:26	49	8:16	36	7:23	23	6:47	10	6:19
	61	10:06	48	8:11	35	7:19	22	6:45	9	6:17
	60	9:50	47	8:06	34	7:16	21	6:43	8	6:15
	59	9:37	46	8:01	33	7:13	20	6:40	7	6:13
	58	9:25	45	7:56	32	7:10	19	6:38	6	6:11
	57	9:14	44	7:52	31	7:08	18	6:36	5	6:09
	56	9:05	43	7:48	30	7:05	17	6:34	4	6:07
	55	8:56	42	7:44	29	7:02	16	6:32	3	6:05
	54	8:48	41	7:40	28	7:00	15	6:30	2	6:03
	53	8:41	40	7:36	27	6:57	14	6:27	1	6:01
	52	8:34	39	7:33	26	6:55	13	6:25	0	6:00
25	65	12:00	51	8:20	37	7:22	23	6:45	9	6:16
	64	10:51	50	8:15	36	7:19	22	6:43	8	6:15
	63	10:24	49	8:09	35	7:16	21	6:41	7	6:13
	62	10:05	48	8:04	34	7:13	20	6:39	6	6:11
	61	9:49	47	8:00	33	7:10	19	6:36	5	6:09
	60	9:35	46	7:55	32	7:07	18	6:34	4	6:07
	59	9:23	45	7:51	31	7:05	17	6:32	3	6:05
	58	9:13	44	7:47	30	7:02	16	6:30	2	6:03
	57	9:03	43	7:43	29	6:59	15	6:28	1	6:01
	56	8:54	42	7:39	28	6:57	14	6:26	0	6:00
	55	8:47	41	7:35	27	6:54	13	6:24		
	54	8:39	40	7:32	26	6:52	12	6:22		
	53	8:32	39	7:28	25	6:50	11	6:20		
	52	8:26	38	7:25	24	6:47	10	6:18		

TIMETABLE B--for stars in the celestial hemisphere OPPOSITE THE CIRCUMPOLAR REGION.

Dec	Time	Dec	Time	Dec	Time	Dec	Time	Dec	Time	Latitude
10	5:38	23	5:07	36	4:29	49	3:29	62	0:00	28
9	5:40	22	5:10	35	4:32	48	3:35	61	1:05	
8	5:42	21	5:12	34	4:35	47	3:40	60	1:31	
7	5:45	20	5:15	33	4:39	46	3:46	59	1:51	
6	5:47	19	5:17	32	4:42	45	3:51	58	2:06	
5	5:49	18	5:20	31	4:45	44	3:56	57	2:20	
4	5:51	17	5:22	30	4:48	43	4:01	56	2:31	
3	5:53	16	5:24	29	4:51	42	4:05	55	2:42	
2	5:55	15	5:27	28	4:54	41	4:09	54	2:51	
1	5:57	14	5:29	27	4:57	40	4:14	53	3:00	
0	6:00	13	5:31	26	4:59	39	4:17	52	3:08	
		12	5:34	25	5:02	38	4:21	51	3:15	
		11	5:36	24	5:05	37	4:25	50	3:22	
11	5:37	24	5:07	37	4:29	50	3:30	63	0:00	27
10	5:39	23	5:10	36	4:33	49	3:36	62	1:06	
9	5:41	22	5:12	35	4:36	48	3:42	61	1:32	
8	5:43	21	5:14	34	4:39	47	3:47	60	1:52	
7	5:45	20	5:17	33	4:42	46	3:52	59	2:08	
6	5:47	19	5:19	32	4:45	45	3:57	58	2:21	
5	5:49	18	5:21	31	4:48	44	4:02	57	2:33	
4	5:51	17	5:24	30	4:51	43	4:06	56	2:43	
3	5:53	16	5:26	29	4:54	42	4:10	55	2:53	
2	5:55	15	5:28	28	4:57	41	4:14	54	3:01	
1	5:57	14	5:30	27	4:59	40	4:18	53	3:09	
0	6:00	13	5:32	26	5:02	39	4:22	52	3:17	
		12	5:35	25	5:05	38	4:26	51	3:24	
12	5:36	25	5:07	38	4:30	51	3:31	64	0:00	26
11	5:38	24	5:09	37	4:33	50	3:37	63	1:07	
10	5:40	23	5:12	36	4:36	49	3:43	62	1:33	
9	5:42	22	5:14	35	4:40	48	3:48	61	1:53	
8	5:44	21	5:16	34	4:43	47	3:53	60	2:09	
7	5:46	20	5:19	33	4:46	46	3:58	59	2:22	
6	5:48	19	5:21	32	4:49	45	4:03	58	2:34	
5	5:50	18	5:23	31	4:51	44	4:07	57	2:45	
4	5:52	17	5:25	30	4:54	43	4:11	56	2:54	
3	5:54	16	5:27	29	4:57	42	4:15	55	3:03	
2	5:56	15	5:29	28	4:59	41	4:19	54	3:11	
1	5:58	14	5:32	27	5:02	40	4:23	53	3:18	
0	6:00	13	5:34	26	5:04	39	4:26	52	3:25	
9	5:43	23	5:14	37	4:37	51	3:39	65	0:00	25
8	5:44	22	5:16	36	4:40	50	3:44	64	1:08	
7	5:46	21	5:18	35	4:43	49	3:50	63	1:35	
6	5:48	20	5:20	34	4:46	48	3:55	62	1:54	
5	5:50	19	5:23	33	4:49	47	3:59	61	2:10	
4	5:52	18	5:25	32	4:52	46	4:04	60	2:24	
3	5:54	17	5:27	31	4:54	45	4:08	59	2:36	
2	5:56	16	5:29	30	4:57	44	4:12	58	2:46	
1	5:58	15	5:31	29	5:00	43	4:16	57	2:56	
0	6:00	14	5:33	28	5:02	42	4:20	56	3:05	
		13	5:35	27	5:05	41	4:24	55	3:12	
		12	5:37	26	5:07	40	4:27	54	3:20	
		11	5:39	25	5:09	39	4:31	53	3:27	
		10	5:41	24	5:12	38	4:34	52	3:33	

TIMETABLE A--for stars in the **SAME CELESTIAL HEMISPHERE** as the circumpolar region.

Latitude	Dec	Time	Dec	Time	Dec	Time	Dec	Time	Dec	Time
24	66	12:00	52	8:18	38	7:21	24	6:45	10	6:18
	65	10:50	51	8:13	37	7:18	23	6:43	9	6:16
	64	10:23	50	8:08	36	7:15	22	6:41	8	6:14
	63	10:03	49	8:03	35	7:12	21	6:39	7	6:12
	62	9:47	48	7:58	34	7:09	20	6:37	6	6:10
	61	9:33	47	7:54	33	7:07	19	6:35	5	6:08
	60	9:21	46	7:49	32	7:04	18	6:33	4	6:07
	59	9:11	45	7:45	31	7:02	17	6:31	3	6:05
	58	9:01	44	7:41	30	6:59	16	6:29	2	6:03
	57	8:53	43	7:38	29	6:57	15	6:27	1	6:01
	56	8:45	42	7:34	28	6:54	14	6:25	0	6:00
	55	8:37	41	7:31	27	6:52	13	6:23		
	54	8:31	40	7:27	26	6:50	12	6:21		
	53	8:24	39	7:24	25	6:47	11	6:19		
23	67	12:00	53	8:17	39	7:20	25	6:45	11	6:18
	66	10:49	52	8:11	38	7:17	24	6:43	10	6:17
	65	10:22	51	8:06	37	7:14	23	6:41	9	6:15
	64	10:01	50	8:01	36	7:11	22	6:39	8	6:13
	63	9:45	49	7:56	35	7:09	21	6:37	7	6:11
	62	9:31	48	7:52	34	7:06	20	6:35	6	6:10
	61	9:19	47	7:48	33	7:04	19	6:33	5	6:08
	60	9:09	46	7:44	32	7:01	18	6:31	4	6:06
	59	8:59	45	7:40	31	6:59	17	6:29	3	6:05
	58	8:51	44	7:36	30	6:56	16	6:27	2	6:03
	57	8:43	43	7:33	29	6:54	15	6:26	1	6:01
	56	8:35	42	7:29	28	6:52	14	6:24	0	6:00
	55	8:29	41	7:26	27	6:49	13	6:22		
	54	8:22	40	7:23	26	6:47	12	6:20		
22	68	12:00	54	8:15	40	7:19	26	6:45	12	6:19
	67	10:48	53	8:09	39	7:16	25	6:43	11	6:18
	66	10:20	52	8:04	38	7:13	24	6:41	10	6:16
	65	10:00	51	7:59	37	7:10	23	6:39	9	6:14
	64	9:43	50	7:55	36	7:08	22	6:37	8	6:13
	63	9:29	49	7:50	35	7:05	21	6:35	7	6:11
	62	9:17	48	7:46	34	7:03	20	6:33	6	6:09
	61	9:07	47	7:42	33	7:00	19	6:31	5	6:08
	60	8:57	46	7:38	32	6:58	18	6:30	4	6:06
	59	8:49	45	7:35	31	6:56	17	6:28	3	6:04
	58	8:41	44	7:31	30	6:53	16	6:26	2	6:03
	57	8:33	43	7:28	29	6:51	15	6:24	1	6:01
	56	8:27	42	7:25	28	6:49	14	6:23	0	6:00
	55	8:20	41	7:22	27	6:47	13	6:21		

TIMETABLE B--for stars in the celestial hemisphere **OPPOSITE THE CIRCUMPOLAR REGION.**

Dec	Time	Dec	Time	Dec	Time	Dec	Time	Dec	Time	Latitude
10	5:41	24	5:14	38	4:38	52	3:41	66	0:00	24
9	5:43	23	5:16	37	4:41	51	3:46	65	1:09	
8	5:45	22	5:18	36	4:44	50	3:51	64	1:36	
7	5:47	21	5:20	35	4:47	49	3:56	63	1:56	
6	5:49	20	5:22	34	4:50	48	4:01	62	2:12	
5	5:51	19	5:24	33	4:52	47	4:05	61	2:26	
4	5:52	18	5:26	32	4:55	46	4:10	60	2:38	
3	5:54	17	5:28	31	4:57	45	4:14	59	2:48	
2	5:56	16	5:30	30	5:00	44	4:18	58	2:58	
1	5:58	15	5:32	29	5:02	43	4:21	57	3:06	
0	6:00	14	5:34	28	5:05	42	4:25	56	3:14	
		13	5:36	27	5:07	41	4:28	55	3:22	
		12	5:38	26	5:09	40	4:32	54	3:28	
		11	5:40	25	5:12	39	4:35	53	3:35	
11	5:41	25	5:14	39	4:39	53	3:42	67	0:00	23
10	5:42	24	5:16	38	4:42	52	3:48	66	1:10	
9	5:44	23	5:18	37	4:45	51	3:53	65	1:37	
8	5:46	22	5:20	36	4:48	50	3:58	64	1:58	
7	5:48	21	5:22	35	4:50	49	4:03	63	2:14	
6	5:49	20	5:24	34	4:53	48	4:07	62	2:28	
5	5:51	19	5:26	33	4:55	47	4:11	61	2:40	
4	5:53	18	5:28	32	4:58	46	4:15	60	2:50	
3	5:54	17	5:30	31	5:00	45	4:19	59	3:00	
2	5:56	16	5:32	30	5:03	44	4:23	58	3:08	
1	5:58	15	5:33	29	5:05	43	4:26	57	3:16	
0	6:00	14	5:35	28	5:07	42	4:30	56	3:24	
		13	5:37	27	5:10	41	4:33	55	3:30	
		12	5:39	26	5:12	40	4:36	54	3:37	
12	5:40	26	5:14	40	4:40	54	3:44	68	0:00	22
11	5:41	25	5:16	39	4:43	53	3:50	67	1:11	
10	5:43	24	5:18	38	4:46	52	3:55	66	1:39	
9	5:45	23	5:20	37	4:49	51	4:00	65	1:59	
8	5:46	22	5:22	36	4:51	50	4:04	64	2:16	
7	5:48	21	5:24	35	4:54	49	4:09	63	2:30	
6	5:50	20	5:26	34	4:56	48	4:13	62	2:42	
5	5:51	19	5:28	33	4:59	47	4:17	61	2:52	
4	5:53	18	5:29	32	5:01	46	4:21	60	3:02	
3	5:55	17	5:31	31	5:03	45	4:24	59	3:10	
2	5:56	16	5:33	30	5:06	44	4:28	58	3:18	
1	5:58	15	5:35	29	5:08	43	4:31	57	3:26	
0	6:00	14	5:36	28	5:10	42	4:34	56	3:32	
		13	5:38	27	5:12	41	4:37	55	3:39	

TIMETABLE A--for stars in the **SAME CELESTIAL HEMISPHERE** as the circumpolar region.

Latitude	Dec	Time	Dec	Time	Dec	Time	Dec	Time	Dec	Time
21	69	12:00	55	8:12	41	7:17	27	6:45	13	6:20
	68	10:47	54	8:07	40	7:15	26	6:43	12	6:18
	67	10:18	53	8:02	39	7:12	25	6:41	11	6:17
	66	9:58	52	7:57	38	7:09	24	6:39	10	6:15
	65	9:41	51	7:53	37	7:07	23	6:37	9	6:13
	64	9:27	50	7:48	36	7:04	22	6:35	8	6:12
	63	9:15	49	7:44	35	7:02	21	6:33	7	6:10
	62	9:04	48	7:40	34	7:00	20	6:32	6	6:09
	61	8:55	47	7:37	33	6:57	19	6:30	5	6:07
	60	8:46	46	7:33	32	6:55	18	6:28	4	6:06
	59	8:38	45	7:30	31	6:53	17	6:26	3	6:04
	58	8:31	44	7:27	30	6:51	16	6:25	2	6:03
	57	8:24	43	7:23	29	6:49	15	6:23	1	6:01
	56	8:18	42	7:20	28	6:47	14	6:21	0	6:00
20	70	12:00	55	8:05	40	7:11	25	6:39	10	6:14
	69	10:45	54	8:00	39	7:08	24	6:37	9	6:13
	68	10:17	53	7:55	38	7:06	23	6:35	8	6:11
	67	9:56	52	7:51	37	7:03	22	6:33	7	6:10
	66	9:39	51	7:46	36	7:01	21	6:32	6	6:08
	65	9:25	50	7:42	35	6:59	20	6:30	5	6:07
	64	9:13	49	7:39	34	6:56	19	6:28	4	6:05
	63	9:02	48	7:35	33	6:54	18	6:27	3	6:04
	62	8:52	47	7:31	32	6:52	17	6:25	2	6:02
	61	8:44	46	7:28	31	6:50	16	6:23	1	6:01
	60	8:36	45	7:25	30	6:48	15	6:22	0	6:00
	59	8:29	44	7:22	29	6:46	14	6:20		
	58	8:22	43	7:19	28	6:44	13	6:19		
	57	8:16	42	7:16	27	6:42	12	6:17		
	56	8:10	41	7:13	26	6:40	11	6:16		
19	71	12:00	56	8:02	41	7:09	26	6:38	11	6:15
	70	10:44	55	7:57	40	7:07	25	6:36	10	6:13
	69	10:15	54	7:53	39	7:04	24	6:35	9	6:12
	68	9:53	53	7:48	38	7:02	23	6:33	8	6:11
	67	9:36	52	7:44	37	7:00	22	6:31	7	6:09
	66	9:22	51	7:40	36	6:57	21	6:30	6	6:08
	65	9:10	50	7:36	35	6:55	20	6:28	5	6:06
	64	8:59	49	7:33	34	6:53	19	6:27	4	6:05
	63	8:50	48	7:29	33	6:51	18	6:25	3	6:04
	62	8:41	47	7:26	32	6:49	17	6:24	2	6:02
	61	8:33	46	7:23	31	6:47	16	6:22	1	6:01
	60	8:26	45	7:20	30	6:45	15	6:21	0	6:00
	59	8:19	44	7:17	29	6:44	14	6:19		
	58	8:13	43	7:14	28	6:42	13	6:18		
	57	8:08	42	7:12	27	6:40	12	6:16		

TIMETABLE B--for stars in the celestial hemisphere **OPPOSITE THE CIRCUMPOLAR REGION.**

Dec	Time	Dec	Time	Dec	Time	Dec	Time	Dec	Time	Latitude
13	5:39	27	5:14	41	4:42	55	3:47	69	0:00	21
12	5:41	26	5:16	40	4:44	54	3:52	68	1:12	
11	5:42	25	5:18	39	4:47	53	3:57	67	1:41	
10	5:44	24	5:20	38	4:50	52	4:02	66	2:01	
9	5:46	23	5:22	37	4:52	51	4:06	65	2:18	
8	5:47	22	5:24	36	4:55	50	4:11	64	2:32	
7	5:49	21	5:26	35	4:57	49	4:15	63	2:44	
6	5:50	20	5:27	34	4:59	48	4:19	62	2:55	
5	5:52	19	5:29	33	5:02	47	4:22	61	3:04	
4	5:53	18	5:31	32	5:04	46	4:26	60	3:13	
3	5:55	17	5:33	31	5:06	45	4:29	59	3:21	
2	5:56	16	5:34	30	5:08	44	4:32	58	3:28	
1	5:58	15	5:36	29	5:10	43	4:36	57	3:35	
0	6:00	14	5:38	28	5:12	42	4:39	56	3:41	
10	5:45	25	5:20	40	4:48	55	3:54	70	0:00	20
9	5:46	24	5:22	39	4:51	54	3:59	69	1:14	
8	5:48	23	5:24	38	4:53	53	4:04	68	1:42	
7	5:49	22	5:26	37	4:56	52	4:08	67	2:03	
6	5:51	21	5:27	36	4:58	51	4:13	66	2:20	
5	5:52	20	5:29	35	5:00	50	4:17	65	2:34	
4	5:54	19	5:31	34	5:03	49	4:20	64	2:46	
3	5:55	18	5:32	33	5:05	48	4:24	63	2:57	
2	5:57	17	5:34	32	5:07	47	4:28	62	3:07	
1	5:58	16	5:36	31	5:09	46	4:31	61	3:15	
0	6:00	15	5:37	30	5:11	45	4:34	60	3:23	
		14	5:39	29	5:13	44	4:37	59	3:30	
		13	5:40	28	5:15	43	4:40	58	3:37	
		12	5:42	27	5:17	42	4:43	57	3:43	
		11	5:43	26	5:19	41	4:46	56	3:49	
11	5:44	26	5:21	41	4:50	56	3:57	71	0:00	19
10	5:46	25	5:23	40	4:52	55	4:02	70	1:15	
9	5:47	24	5:24	39	4:55	54	4:06	69	1:44	
8	5:48	23	5:26	38	4:57	53	4:11	68	2:06	
7	5:50	22	5:28	37	4:59	52	4:15	67	2:23	
6	5:51	21	5:29	36	5:02	51	4:19	66	2:37	
5	5:53	20	5:31	35	5:04	50	4:23	65	2:49	
4	5:54	19	5:32	34	5:06	49	4:26	64	3:00	
3	5:55	18	5:34	33	5:08	48	4:30	63	3:09	
2	5:57	17	5:35	32	5:10	47	4:33	62	3:18	
1	5:58	16	5:37	31	5:12	46	4:36	61	3:26	
0	6:00	15	5:38	30	5:14	45	4:39	60	3:33	
		14	5:40	29	5:15	44	4:42	59	3:40	
		13	5:41	28	5:17	43	4:45	58	3:46	
		12	5:43	27	5:19	42	4:47	57	3:51	

TIMETABLE A--for stars in the **SAME CELESTIAL HEMISPHERE** as the circumpolar region.

Latitude	Dec	Time	Dec	Time	Dec	Time	Dec	Time	Dec	Time
18	72	12:00	57	8:00	42	7:08	27	6:38	12	6:15
	71	10:42	56	7:55	41	7:05	26	6:36	11	6:14
	70	10:12	55	7:50	40	7:03	25	6:34	10	6:13
	69	9:51	54	7:46	39	7:01	24	6:33	9	6:11
	68	9:34	53	7:42	38	6:58	23	6:31	8	6:10
	67	9:19	52	7:38	37	6:56	22	6:30	7	6:09
	66	9:07	51	7:34	36	6:54	21	6:28	6	6:07
	65	8:56	50	7:31	35	6:52	20	6:27	5	6:06
	64	8:47	49	7:27	34	6:50	19	6:25	4	6:05
	63	8:38	48	7:24	33	6:48	18	6:24	3	6:03
	62	8:30	47	7:21	32	6:46	17	6:22	2	6:02
	61	8:23	46	7:18	31	6:45	16	6:21	1	6:01
	60	8:16	45	7:15	30	6:43	15	6:19	0	6:00
	59	8:10	44	7:13	29	6:41	14	6:18		
	58	8:05	43	7:10	28	6:39	13	6:17		
17	73	12:00	58	7:57	43	7:06	28	6:37	13	6:16
	72	10:40	57	7:52	42	7:03	27	6:35	12	6:14
	71	10:10	56	7:47	41	7:01	26	6:34	11	6:13
	70	9:48	55	7:43	40	6:59	25	6:32	10	6:12
	69	9:31	54	7:39	39	6:57	24	6:31	9	6:11
	68	9:16	53	7:35	38	6:55	23	6:29	8	6:09
	67	9:04	52	7:32	37	6:53	22	6:28	7	6:08
	66	8:53	51	7:28	36	6:51	21	6:26	6	6:07
	65	8:43	50	7:25	35	6:49	20	6:25	5	6:06
	64	8:35	49	7:22	34	6:47	19	6:24	4	6:04
	63	8:27	48	7:19	33	6:45	18	6:22	3	6:03
	62	8:20	47	7:16	32	6:44	17	6:21	2	6:02
	61	8:13	46	7:13	31	6:42	16	6:20	1	6:01
	60	8:07	45	7:11	30	6:40	15	6:18	0	6:00
	59	8:02	44	7:08	29	6:39	14	6:17		
16	74	12:00	59	7:54	44	7:04	29	6:36	14	6:16
	73	10:38	58	7:49	43	7:02	28	6:35	13	6:15
	72	10:07	57	7:44	42	6:59	27	6:33	12	6:13
	71	9:45	56	7:40	41	6:57	26	6:32	11	6:12
	70	9:27	55	7:36	40	6:55	25	6:30	10	6:11
	69	9:13	54	7:32	39	6:53	24	6:29	9	6:10
	68	9:00	53	7:29	38	6:51	23	6:27	8	6:09
	67	8:49	52	7:26	37	6:49	22	6:26	7	6:08
	66	8:40	51	7:22	36	6:48	21	6:25	6	6:06
	65	8:31	50	7:19	35	6:46	20	6:23	5	6:05
	64	8:24	49	7:17	34	6:44	19	6:22	4	6:04
	63	8:16	48	7:14	33	6:42	18	6:21	3	6:03
	62	8:10	47	7:11	32	6:41	17	6:20	2	6:02
	61	8:04	46	7:09	31	6:39	16	6:18	1	6:01
	60	7:59	45	7:06	30	6:38	15	6:17	0	6:00

TIMETABLE B--for stars in the celestial hemisphere **OPPOSITE THE CIRCUMPOLAR REGION.**

Dec	Time	Dec	Time	Dec	Time	Dec	Time	Dec	Time	Latitude
12	5:44	27	5:21	42	4:51	57	3:59	72	0:00	18
11	5:45	26	5:23	41	4:54	56	4:04	71	1:17	
10	5:46	25	5:25	40	4:56	55	4:09	70	1:47	
9	5:48	24	5:26	39	4:58	54	4:13	69	2:08	
8	5:49	23	5:28	38	5:01	53	4:17	68	2:25	
7	5:50	22	5:29	37	5:03	52	4:21	67	2:40	
6	5:52	21	5:31	36	5:05	51	4:25	66	2:52	
5	5:53	20	5:32	35	5:07	50	4:28	65	3:03	
4	5:54	19	5:34	34	5:09	49	4:32	64	3:12	
3	5:56	18	5:35	33	5:11	48	4:35	63	3:21	
2	5:57	17	5:37	32	5:13	47	4:38	62	3:29	
1	5:58	16	5:38	31	5:14	46	4:41	61	3:36	
0	6:00	15	5:40	30	5:16	45	4:44	60	3:43	
		14	5:41	29	5:18	44	4:46	59	3:49	
		13	5:42	28	5:20	43	4:49	58	3:54	
13	5:43	28	5:22	43	4:53	58	4:02	73	0:00	17
12	5:45	27	5:24	42	4:56	57	4:07	72	1:19	
11	5:46	26	5:25	41	4:58	56	4:12	71	1:49	
10	5:47	25	5:27	40	5:00	55	4:16	70	2:11	
9	5:48	24	5:28	39	5:02	54	4:20	69	2:28	
8	5:50	23	5:30	38	5:04	53	4:24	68	2:43	
7	5:51	22	5:31	37	5:06	52	4:27	67	2:55	
6	5:52	21	5:33	36	5:08	51	4:31	66	3:06	
5	5:53	20	5:34	35	5:10	50	4:34	65	3:16	
4	5:55	19	5:35	34	5:12	49	4:37	64	3:24	
3	5:56	18	5:37	33	5:14	48	4:40	63	3:32	
2	5:57	17	5:38	32	5:15	47	4:43	62	3:39	
1	5:58	16	5:39	31	5:17	46	4:46	61	3:46	
0	6:00	15	5:41	30	5:19	45	4:48	60	3:52	
		14	5:42	29	5:20	44	4:51	59	3:57	
14	5:43	29	5:23	44	4:55	59	4:05	74	0:00	16
13	5:44	28	5:24	43	4:57	58	4:10	73	1:21	
12	5:46	27	5:26	42	5:00	57	4:15	72	1:52	
11	5:47	26	5:27	41	5:02	56	4:19	71	2:14	
10	5:48	25	5:29	40	5:04	55	4:23	70	2:32	
9	5:49	24	5:30	39	5:06	54	4:27	69	2:46	
8	5:50	23	5:32	38	5:08	53	4:30	68	2:59	
7	5:51	22	5:33	37	5:10	52	4:33	67	3:10	
6	5:53	21	5:34	36	5:11	51	4:37	66	3:19	
5	5:54	20	5:36	35	5:13	50	4:40	65	3:28	
4	5:55	19	5:37	34	5:15	49	4:42	64	3:35	
3	5:56	18	5:38	33	5:17	48	4:45	63	3:43	
2	5:57	17	5:39	32	5:18	47	4:48	62	3:49	
1	5:58	16	5:41	31	5:20	46	4:50	61	3:55	
0	6:00	15	5:42	30	5:21	45	4:53	60	4:00	

TIMETABLE A--for stars in the **SAME CELESTIAL HEMISPHERE** as the circumpolar region.

Latitude	Dec	Time	Dec	Time	Dec	Time	Dec	Time	Dec	Time
15	75	12:00	59	7:45	43	6:57	27	6:31	11	6:11
	74	10:36	58	7:41	42	6:55	26	6:30	10	6:10
	73	10:04	57	7:37	41	6:53	25	6:28	9	6:09
	72	9:42	56	7:33	40	6:51	24	6:27	8	6:08
	71	9:24	55	7:29	39	6:50	23	6:26	7	6:07
	70	9:09	54	7:26	38	6:48	22	6:24	6	6:06
	69	8:57	53	7:23	37	6:46	21	6:23	5	6:05
	68	8:46	52	7:20	36	6:44	20	6:22	4	6:04
	67	8:36	51	7:17	35	6:43	19	6:21	3	6:03
	66	8:28	50	7:14	34	6:41	18	6:19	2	6:02
	65	8:20	49	7:11	33	6:40	17	6:18	1	6:01
	64	8:13	48	7:09	32	6:38	16	6:17	0	6:00
	63	8:06	47	7:06	31	6:37	15	6:16		
	62	8:01	46	7:04	30	6:35	14	6:15		
	61	7:55	45	7;02	29	6:34	13	6:14		
	60	7:50	44	6:59	28	6:32	12	6:13		
14	76	12:00	60	7:42	44	6:55	28	6:30	12	6:12
	75	10:34	59	7:38	43	6:53	27	6:29	11	6:11
	74	10:01	58	7:34	42	6:51	26	6:27	10	6:10
	73	9:38	57	7:30	41	6:50	25	6:26	9	6:09
	72	9:20	56	7:26	40	6:48	24	6:25	8	6:08
	71	9:05	55	7:23	39	6:46	23	6:24	7	6:07
	70	8:52	54	7:20	38	6:44	22	6:23	6	6:06
	69	8:42	53	7:17	37	6:43	21	6:21	5	6:04
	68	8:32	52	7:14	36	6:41	20	6:20	4	6:03
	67	8:23	51	7:11	35	6:40	19	6:19	3	6:02
	66	8:16	50	7:09	34	6:38	18	6:18	2	6:01
	65	8:09	49	7:06	33	6:37	17	6:17	1	6:00
	64	8:02	48	7:04	32	6:35	16	6:16	0	6:00
	63	7:57	47	7:02	31	6:34	15	6:15		
	62	7:51	46	6:59	30	6:33	14	6:14		
	61	7:46	45	6:57	29	6:31	13	6:13		
13	77	12:00	61	7:38	45	6:53	29	6:29	13	6:12
	76	10:31	60	7:34	44	6:51	28	6:28	12	6:11
	75	9:57	59	7:30	43	6:49	27	6:27	11	6:10
	74	9:34	58	7:26	42	6:47	26	6:25	10	6:09
	73	9:16	57	7:23	41	6:46	25	6:24	9	6:08
	72	9:01	56	7:20	40	6:44	24	6:23	8	6:07
	71	8:48	55	7:17	39	6:43	23	6:22	7	6:06
	70	8:37	54	7:14	38	6:41	22	6:21	6	6:05
	69	8:27	53	7:11	37	6:40	21	6:20	5	6:04
	68	8:19	52	7:08	36	6:38	20	6:19	4	6:03
	67	8:11	51	7:06	35	6:37	19	6:18	3	6:02
	66	8:04	50	7:03	34	6:35	18	6:17	2	6:01
	65	7:58	49	7:01	33	6:34	17	6:16	1	6:00
	64	7:53	48	6:59	32	6:33	16	6:15	0	6:00
	63	7:47	47	6:57	31	6:31	15	6:14		
	62	7:42	46	6:55	30	6:30	14	6:13		

TIMETABLE B--for stars in the celestial hemisphere OPPOSITE THE CIRCUMPOLAR REGION.

Dec	Time	Dec	Time	Dec	Time	Dec	Time	Dec	Time	Latitude
11	5:48	27	5:28	43	5:02	59	4:14	75	0:00	15
10	5:49	26	5:29	42	5:04	58	4:18	74	1:23	
9	5:50	25	5:31	41	5:06	57	4:22	73	1:55	
8	5:51	24	5:32	40	5:08	56	4:26	72	2:17	
7	5:52	23	5:33	39	5:09	55	4:30	71	2:35	
6	5:53	22	5:35	38	5:11	54	4:33	70	2:50	
5	5:54	21	5:36	37	5:13	53	4:36	69	3:02	
4	5:55	20	5:37	36	5:15	52	4:39	68	3:13	
3	5:56	19	5:38	35	5:16	51	4:42	67	3:23	
2	5:57	18	5:40	34	5:18	50	4:45	66	3:31	
1	5:58	17	5:41	33	5:19	49	4:48	65	3:39	
0	6:00	16	5:42	32	5:21	48	4:50	64	3:46	
		15	5:43	31	5:22	47	4:53	63	3:53	
		14	5:44	30	5:24	46	4:55	62	3:58	
		13	5:45	29	5:25	45	4:57	61	4:04	
		12	5:46	28	5:27	44	5:00	60	4:09	
12	5:47	28	5:29	44	5:04	60	4:17	76	0:00	14
11	5:48	27	5:30	43	5:06	59	4:21	75	1:25	
10	5:49	26	5:32	42	5:08	58	4:25	74	1:58	
9	5:50	25	5:33	41	5:09	57	4:29	73	2:21	
8	5:51	24	5:34	40	5:11	56	4:33	72	2:39	
7	5:52	23	5:35	39	5:13	55	4:36	71	2:54	
6	5:53	22	5:36	38	5:15	54	4:39	70	3:07	
5	5:55	21	5:38	37	5:16	53	4:42	69	3:17	
4	5:56	20	5:39	36	5:18	52	4:45	68	3:27	
3	5:57	19	5:40	35	5:19	51	4:48	67	3:36	
2	5:58	18	5:41	34	5:21	50	4:50	66	3:43	
1	5:59	17	5:42	33	5:22	49	4:53	65	3:50	
0	6:00	16	5:43	32	5:24	48	4:55	64	3:57	
		15	5:44	31	5:25	47	4:57	63	4:02	
		14	5:45	30	5:26	46	5:00	62	4:08	
		13	5:46	29	5:28	45	5:02	61	4:13	
13	5:47	29	5:30	45	5:06	61	4:21	77	0:00	13
12	5:48	28	5:31	44	5:08	60	4:25	76	1:28	
11	5:49	27	5:32	43	5:10	59	4:29	75	2:02	
10	5:50	26	5:34	42	5:12	58	4:33	74	2:25	
9	5:51	25	5:35	41	5:13	57	4:36	73	2:43	
8	5:52	24	5:36	40	5:15	56	4:39	72	2:58	
7	5:53	23	5:37	39	5:16	55	4:42	71	3:11	
6	5:54	22	5:38	38	5:18	54	4:45	70	3:22	
5	5:55	21	5:39	37	5:19	53	4:48	69	3:32	
4	5:56	20	5:40	36	5:21	52	4:51	68	3:40	
3	5:57	19	5:41	35	5:22	51	4:53	67	3:48	
2	5:58	18	5:42	34	5:24	50	4:56	66	3:55	
1	5:59	17	5:43	33	5:25	49	4:58	65	4:01	
0	6:00	16	5:44	32	5:26	48	5:00	64	4:06	
		15	5:45	31	5:28	47	5:02	63	4:12	
		14	5:46	30	5:29	46	5:04	62	4:17	

TIMETABLE A--for stars in the SAME CELESTIAL HEMISPHERE as the circumpolar region.

Latitude	Dec	Time	Dec	Time	Dec	Time	Dec	Time	Dec	Time
12	78	12:00	62	7:34	46	6:50	30	6:28	14	6:12
	77	10:28	61	7:30	45	6:49	29	6:27	13	6:11
	76	9:53	60	7:26	44	6:47	28	6:25	12	6:10
	75	9:29	59	7:22	43	6:45	27	6:24	11	6:09
	74	9:11	58	7:19	42	6:44	26	6:23	10	6:08
	73	8:56	57	7:16	41	6:42	25	6:22	9	6:07
	72	8:43	56	7:13	40	6:41	24	6:21	8	6:06
	71	8:32	55	7:10	39	6:39	23	6:20	7	6:05
	70	8:22	54	7:08	38	6:38	22	6:19	6	6:05
	69	8:14	53	7:05	37	6:36	21	6:18	5	6:04
	68	8:06	52	7:03	36	6:35	20	6:17	4	6:03
	67	8:00	51	7:00	35	6:34	19	6:16	3	6:02
	66	7:54	50	6:58	34	6:32	18	6:15	2	6:01
	65	7:48	49	6:56	33	6:31	17	6:14	1	6:00
	64	7:43	48	6:54	32	6:30	16	6:13	0	6:00
	63	7:38	47	6:52	31	6:29	15	6:13		
11	79	12:00	63	7:29	47	6:48	31	6:26	15	6:11
	78	10:24	62	7:25	46	6:46	30	6:25	14	6:11
	77	9:49	61	7:22	45	6:44	29	6:24	13	6:10
	76	9:24	60	7:18	44	6:43	28	6:23	12	6:09
	75	9:06	59	7:15	43	6:41	27	6:22	11	6:08
	74	8:50	58	7:12	42	6:40	26	6:21	10	6:07
	73	8:37	57	7:09	41	6:38	25	6:20	9	6:07
	72	8:26	56	7:06	40	6:37	24	6:19	8	6:06
	71	8:17	55	7:04	39	6:36	23	6:18	7	6:05
	70	8:09	54	7:02	38	6:34	22	6:18	6	6:04
	69	8:01	53	6:59	37	6:33	21	6:17	5	6:03
	68	7:55	52	6:57	36	6:32	20	6:16	4	6:03
	67	7:49	51	6:55	35	6:31	19	6:15	3	6:02
	66	7:43	50	6:53	34	6:30	18	6:14	2	6:01
	65	7:38	49	6:51	33	6:29	17	6:13	1	6:00
	64	7:33	48	6:49	32	6:27	16	6:12	0	6:00
10	80	12:00	63	7:20	46	6:42	29	6:22	12	6:08
	79	10:20	62	7:17	45	6:40	28	6:21	11	6:07
	78	9:44	61	7:14	44	6:39	27	6:20	10	6:07
	77	9:19	60	7:11	43	6:37	26	6:19	9	6:06
	76	9:00	59	7:08	42	6:36	25	6:18	8	6:05
	75	8:44	58	7:05	41	6:35	24	6:18	7	6:04
	74	8:31	57	7:03	40	6:34	23	6:17	6	6:04
	73	8:20	56	7:00	39	6:32	22	6:16	5	6:03
	72	8:11	55	6:58	38	6:31	21	6:15	4	6:02
	71	8:03	54	6:56	37	6:30	20	6:14	3	6:02
	70	7:55	53	6:54	36	6:29	19	6:13	2	6:01
	69	7:49	52	6:52	35	6:28	18	6:13	1	6:00
	68	7:43	51	6:50	34	6:27	17	6:12	0	6:00
	67	7:38	50	6:48	33	6:26	16	6:11		
	66	7:33	49	6:46	32	6:25	15	6:10		
	65	7:28	48	6:45	31	6:24	14	6:10		
	64	7:24	47	6:43	30	6:23	13	6:09		

TIMETABLE B--for stars in the celestial hemisphere **OPPOSITE THE CIRCUMPOLAR REGION.**

Dec	Time	Dec	Time	Dec	Time	Dec	Time	Dec	Time	Latitude
14	5:47	30	5:31	46	5:09	62	4:25	78	0:00	12
13	5:48	29	5:32	45	5:10	61	4:29	77	1:31	
12	5:49	28	5:34	44	5:12	60	4:33	76	2:06	
11	5:50	27	5:35	43	5:14	59	4:37	75	2:30	
10	5:51	26	5:36	42	5:15	58	4:40	74	2:48	
9	5:52	25	5:37	41	5:17	57	4:43	73	3:03	
8	5:53	24	5:38	40	5:18	56	4:46	72	3:16	
7	5:54	23	5:39	39	5:20	55	4:49	71	3:27	
6	5:54	22	5:40	38	5:21	54	4:51	70	3:37	
5	5:55	21	5:41	37	5:23	53	4:54	69	3:45	
4	5:56	20	5:42	36	5:24	52	4:56	68	3:53	
3	5:57	19	5:43	35	5:25	51	4:59	67	3:59	
2	5:58	18	5:44	34	5:27	50	5:01	66	4:05	
1	5:59	17	5:45	33	5:28	49	5:03	65	4:11	
0	6:00	16	5:46	32	5:29	48	5:05	64	4:16	
		15	5:46	31	5:30	47	5:07	63	4:21	
15	5:48	31	5:33	47	5:11	63	4:30	79	0:00	11
14	5:48	30	5:34	46	5:13	62	4:34	78	1:35	
13	5:49	29	5:35	45	5:15	61	4:37	77	2:10	
12	5:50	28	5:36	44	5:16	60	4:41	76	2:35	
11	5:51	27	5:37	43	5:18	59	4:44	75	2:53	
10	5:52	26	5:38	42	5:19	58	4:47	74	3:09	
9	5:52	25	5:39	41	5:21	57	4:50	73	3:22	
8	5:53	24	5:40	40	5:22	56	4:53	72	3:33	
7	5:54	23	5:41	39	5:23	55	4:55	71	3:42	
6	5:55	22	5:41	38	5:25	54	4:57	70	3:50	
5	5:56	21	5:42	37	5:26	53	5:00	69	3:58	
4	5:56	20	5:43	36	5:27	52	5:02	68	4:04	
3	5:57	19	5:44	35	5:28	51	5:04	67	4:10	
2	5:58	18	5:45	34	5:29	50	5:06	66	4:16	
1	5:59	17	5:46	33	5:30	49	5:08	65	4:21	
0	6:00	16	5:47	32	5:32	48	5:10	64	4:26	
12	5:51	29	5:37	46	5:17	63	4:39	80	0:00	10
11	5:52	28	5:38	45	5:19	62	4:42	79	1:39	
10	5:52	27	5:39	44	5:20	61	4:45	78	2:15	
9	5:53	26	5:40	43	5:22	60	4:48	77	2:40	
8	5:54	25	5:41	42	5:23	59	4:51	76	2:59	
7	5:55	24	5:41	41	5:24	58	4:54	75	3:15	
6	5:55	23	5:42	40	5:25	57	4:56	74	3:28	
5	5:56	22	5:43	39	5:27	56	4:59	73	3:39	
4	5:57	21	5:44	38	5:28	55	5:01	72	3:48	
3	5:57	20	5:45	37	5:29	54	5:03	71	3:56	
2	5:58	19	5:46	36	5:30	53	5:05	70	4:04	
1	5:59	18	5:46	35	5:31	52	5:07	69	4:10	
0	6:00	17	5:47	34	5:32	51	5:09	68	4:16	
		16	5:48	33	5:33	50	5:11	67	4:21	
		15	5:49	32	5:34	49	5:13	66	4:26	
		14	5:49	31	5:35	48	5:14	65	4:31	
		13	5:50	30	5:36	47	5:16	64	4:35	

TIMETABLE A--for stars in the SAME CELESTIAL HEMISPHERE as the circumpolar region.

Latitude	Dec	Time	Dec	Time	Dec	Time	Dec	Time	Dec	Time
9	81	12:00	64	7:15	47	6:39	30	6:20	13	6:08
	80	10:15	63	7:12	46	6:37	29	6:20	12	6:07
	79	9:38	62	7:09	45	6:36	28	6:19	11	6:07
	78	9:12	61	7:06	44	6:35	27	6:18	10	6:06
	77	8:53	60	7:03	43	6:33	26	6:17	9	6:05
	76	8:37	59	7:01	42	6:32	25	6:16	8	6:05
	75	8:24	58	6:58	41	6:31	24	6:16	7	6:04
	74	8:14	57	6:56	40	6:30	23	6:15	6	6:03
	73	8:04	56	6:54	39	6:29	22	6:14	5	6:03
	72	7:56	55	6:52	38	6:28	21	6:13	4	6:02
	71	7:49	54	6:50	37	6:27	20	6:13	3	6:01
	70	7:43	53	6:48	36	6:26	19	6:12	2	6:01
	69	7:37	52	6:46	35	6:25	18	6:11	1	6:00
	68	7:32	51	6:45	34	6:24	17	6:11	0	6:00
	67	7:27	50	6:43	33	6:23	16	6:10		
	66	7:23	49	6:41	32	6:22	15	6:09		
	65	7:19	48	6:40	31	6:21	14	6:09		
8	82	12:00	65	7:10	48	6:35	31	6:19	14	6:08
	81	10:10	64	7:06	47	6:34	30	6:18	13	6:07
	80	9:31	63	7:04	46	6:33	29	6:17	12	6:06
	79	9:05	62	7:01	45	6:32	28	6:17	11	6:06
	78	8:45	61	6:58	44	6:31	27	6:16	10	6:05
	77	8:29	60	6:56	43	6:30	26	6:15	9	6:05
	76	8:17	59	6:54	42	6:29	25	6:15	8	6:04
	75	8:06	58	6:51	41	6:28	24	6:14	7	6:03
	74	7:57	57	6:49	40	6:27	23	6:13	6	6:03
	73	7:49	56	6:48	39	6:26	22	6:13	5	6:02
	72	7:42	55	6:46	38	6:25	21	6:12	4	6:02
	71	7:36	54	6:44	37	6:24	20	6:11	3	6:01
	70	7:30	53	6:42	36	6:23	19	6:11	2	6:01
	69	7:25	52	6:41	35	6:22	18	6:10	1	6:00
	68	7:21	51	6:39	34	6:21	17	6:09	0	6:00
	67	7:17	50	6:38	33	6:20	16	6:09		
	66	7:13	49	6:37	32	6:20	15	6:08		
7	83	12:00	66	7:04	49	6:32	32	6:17	15	6:07
	82	10:03	65	7:01	48	6:31	31	6:16	14	6:07
	81	9:23	64	6:58	47	6:30	30	6:16	13	6:06
	80	8:56	63	6:55	46	6:29	29	6:15	12	6:05
	79	8:36	62	6:53	45	6:28	28	6:14	11	6:05
	78	8:21	61	6:51	44	6:27	27	6:14	10	6:04
	77	8:08	60	6:49	43	6:26	26	6:13	9	6:04
	76	7:58	59	6:47	42	6:25	25	6:13	8	6:03
	75	7:49	58	6:45	41	6:24	24	6:12	7	6:03
	74	7:41	57	6:43	40	6:23	23	6:11	6	6:02
	73	7:34	56	6:41	39	6:22	22	6:11	5	6:02
	72	7:28	55	6:40	38	6:22	21	6:10	4	6:01
	71	7:23	54	6:38	37	6:21	20	6:10	3	6:01
	70	7:18	53	6:37	36	6:20	19	6:09	2	6:00
	69	7:14	52	6:36	35	6:19	18	6:09	1	6:00
	68	7:10	51	6:34	34	6:19	17	6:08	0	6:00
	67	7:07	50	6:33	33	6:18	16	6:08		

TIMETABLE B--for stars in the celestial hemisphere OPPOSITE THE CIRCUMPOLAR REGION.

Dec	Time	Dec	Time	Dec	Time	Dec	Time	Dec	Time	Latitude
13	5:51	30	5:39	47	5:20	64	4:44	81	0:00	9
12	5:52	29	5:39	46	5:22	63	4:47	80	1:44	
11	5:52	28	5:40	45	5:23	62	4:50	79	2:21	
10	5:53	27	5:41	44	5:24	61	4:53	78	2:47	
9	5:54	26	5:42	43	5:26	60	4:56	77	3:06	
8	5:54	25	5:43	42	5:27	59	4:58	76	3:22	
7	5:55	24	5:43	41	5:28	58	5:01	75	3:35	
6	5:56	23	5:44	40	5:29	57	5:03	74	3:45	
5	5:56	22	5:45	39	5:30	56	5:05	73	3:55	
4	5:57	21	5:46	38	5:31	55	5:07	72	4:03	
3	5:58	20	5:46	37	5:32	54	5:09	71	4:10	
2	5:58	19	5:47	36	5:33	53	5:11	70	4:16	
1	5:59	18	5:48	35	5:34	52	5:13	69	4:22	
0	6:00	17	5:48	34	5:35	51	5:14	68	4:27	
		16	5:49	33	5:36	50	5:16	67	4:32	
		15	5:50	32	5:37	49	5:18	66	4:36	
		14	5:50	31	5:38	48	5:19	65	4:40	
14	5:51	31	5:40	48	5:24	65	4:49	82	0:00	8
13	5:52	30	5:41	47	5:25	64	4:53	81	1:49	
12	5:53	29	5:42	46	5:26	63	4:55	80	2:28	
11	5:53	28	5:42	45	5:27	62	4:58	79	2:54	
10	5:54	27	5:43	44	5:28	61	5:01	78	3:14	
9	5:54	26	5:44	43	5:29	60	5:03	77	3:30	
8	5:55	25	5:44	42	5:30	59	5:05	76	3:42	
7	5:56	24	5:45	41	5:31	58	5:08	75	3:53	
6	5:56	23	5:46	40	5:32	57	5:10	74	4:02	
5	5:57	22	5:46	39	5:33	56	5:11	73	4:10	
4	5:57	21	5:47	38	5:34	55	5:13	72	4:17	
3	5:58	20	5:48	37	5:35	54	5:15	71	4:23	
2	5:58	19	5:48	36	5:36	53	5:17	70	4:29	
1	5:59	18	5:49	35	5:37	52	5:18	69	4:34	
0	6:00	17	5:50	34	5:38	51	5:20	68	4:38	
		16	5:50	33	5:39	50	5:21	67	4:42	
		15	5:51	32	5:39	49	5:22	66	4:46	
15	5:52	32	5:42	49	5:27	66	4:55	83	0:00	7
14	5:52	31	5:43	48	5:28	65	4:58	82	1:56	
13	5:53	30	5:43	47	5:29	64	5:01	81	2:36	
12	5:54	29	5:44	46	5:30	63	5:04	80	3:03	
11	5:54	28	5:45	45	5:31	62	5:06	79	3:23	
10	5:55	27	5:45	44	5:32	61	5:08	78	3:38	
9	5:55	26	5:46	43	5:33	60	5:10	77	3:51	
8	5:56	25	5:46	42	5:34	59	5:12	76	4:01	
7	5:56	24	5:47	41	5:35	58	5:14	75	4:10	
6	5:57	23	5:48	40	5:36	57	5:16	74	4:18	
5	5:57	22	5:48	39	5:37	56	5:18	73	4:25	
4	5:58	21	5:49	38	5:37	55	5:19	72	4:31	
3	5:58	20	5:49	37	5:38	54	5:21	71	4:36	
2	5:59	19	5:50	36	5:39	53	5:22	70	4:41	
1	5:59	18	5:50	35	5:40	52	5:23	69	4:45	
0	6:00	17	5:51	34	5:40	51	5:25	68	4:49	
		16	5:51	33	5:41	50	5:26	67	4:52	

TIMETABLE A--for stars in the **SAME CELESTIAL HEMISPHERE** as the circumpolar region.

Latitude	Dec	Time	Dec	Time	Dec	Time	Dec	Time	Dec	Time
6	84	12:00	67	6:57	50	6:28	33	6:15	16	6:06
	83	9:55	66	6:54	49	6:27	32	6:15	15	6:06
	82	9:13	65	6:52	48	6:26	31	6:14	14	6:06
	81	8:46	64	6:49	47	6:25	30	6:13	13	6:05
	80	8:26	63	6:47	46	6:24	29	6:13	12	6:05
	79	8:10	62	6:45	45	6:24	28	6:12	11	6:04
	78	7:58	61	6:43	44	6:23	27	6:12	10	6:04
	77	7:48	60	6:41	43	6:22	26	6:11	9	6:03
	76	7:39	59	6:40	42	6:21	25	6:11	8	6:03
	75	7:32	58	6:38	41	6:20	24	6:10	7	6:02
	74	7:26	57	6:37	40	6:20	23	6:10	6	6:02
	73	7:20	56	6:35	39	6:19	22	6:09	5	6:02
	72	7:15	55	6:34	38	6:18	21	6:09	4	6:01
	71	7:11	54	6:33	37	6:18	20	6:08	3	6:01
	70	7:07	53	6:32	36	6:17	19	6:08	2	6:00
	69	7:03	52	6:30	35	6:16	18	6:07	1	6:00
	68	7:00	51	6:29	34	6:16	17	6:07	0	6:00
5	85	12:00	67	6:47	49	6:23	31	6:12	13	6:04
	84	9:45	66	6:45	48	6:22	30	6:11	12	6:04
	83	9:01	65	6:43	47	6:21	29	6:11	11	6:03
	82	8:34	64	6:41	46	6:20	28	6:10	10	6:03
	81	8:14	63	6:39	45	6:20	27	6:10	9	6:03
	80	7:58	62	6:37	44	6:19	26	6:09	8	6:02
	79	7:46	61	6:36	43	6:18	25	6:09	7	6:02
	78	7:37	60	6:34	42	6:18	24	6:08	6	6:02
	77	7:29	59	6:33	41	6:17	23	6:08	5	6:01
	76	7:22	58	6:32	40	6:16	22	6:08	4	6:01
	75	7:16	57	6:30	39	6:16	21	6:07	3	6:01
	74	7:11	56	6:29	38	6:15	20	6:07	2	6:00
	73	7:06	55	6:28	37	6:15	19	6:06	1	6:00
	72	7:02	54	6:27	36	6:14	18	6:06	0	6:00
	71	6:58	53	6:26	35	6:14	17	6:06		
	70	6:55	52	6:25	34	6:13	16	6:05		
	69	6:52	51	6:24	33	6:13	15	6:05		
	68	6:50	50	6:23	32	6:12	14	6:04		
4	86	12:00	68	6:39	50	6:19	32	6:10	14	6:03
	85	9:32	67	6:37	49	6:18	31	6:09	13	6:03
	84	8:46	66	6:36	48	6:17	30	6:09	12	6:03
	83	8:18	65	6:34	47	6:17	29	6:08	11	6:03
	82	7:59	64	6:32	46	6:16	28	6:08	10	6:02
	81	7:44	63	6:31	45	6:16	27	6:08	9	6:02
	80	7:33	62	6:30	44	6:15	26	6:07	8	6:02
	79	7:24	61	6:28	43	6:14	25	6:07	7	6:01
	78	7:16	60	6:27	42	6:14	24	6:07	6	6:01
	77	7:10	59	6:26	41	6:13	23	6:06	5	6:01
	76	7:05	58	6:25	40	6:13	22	6:06	4	6:01
	75	7:00	57	6:24	39	6:12	21	6:06	3	6:00
	74	6:56	56	6:23	38	6:12	20	6:05	2	6:00
	73	6:52	55	6:22	37	6:12	19	6:05	1	6:00
	72	6:49	54	6:22	36	6:11	18	6:05	0	6:00
	71	6:46	53	6:21	35	6:11	17	6:04		
	70	6:44	52	6:20	34	6:10	16	6:04		
	69	6:41	51	6:19	33	6:10	15	6:04		

TIMETABLE B--for stars in the celestial hemisphere **OPPOSITE THE CIRCUMPOLAR REGION.**

Dec	Time	Dec	Time	Dec	Time	Dec	Time	Dec	Time	Latitude
16	5:53	33	5:44	50	5:31	67	5:02	84	0:00	6
15	5:53	32	5:44	49	5:32	66	5:05	83	2:04	
14	5:53	31	5:45	48	5:33	65	5:07	82	2:46	
13	5:54	30	5:46	47	5:34	64	5:10	81	3:13	
12	5:54	29	5:46	46	5:35	63	5:12	80	3:33	
11	5:55	28	5:47	45	5:35	62	5:14	79	3:49	
10	5:55	27	5:47	44	5:36	61	5:16	78	4:01	
9	5:56	26	5:48	43	5:37	60	5:18	77	4:11	
8	5:56	25	5:48	42	5:38	59	5:19	76	4:20	
7	5:57	24	5:49	41	5:39	58	5:21	75	4:27	
6	5:57	23	5:49	40	5:39	57	5:22	74	4:33	
5	5:57	22	5:50	39	5:40	56	5:24	73	4:39	
4	5:58	21	5:50	38	5:41	55	5:25	72	4:44	
3	5:58	20	5:51	37	5:41	54	5:26	71	4:48	
2	5:59	19	5:51	36	5:42	53	5:27	70	4:52	
1	5:59	18	5:52	35	5:43	52	5:29	69	4:56	
0	6:00	17	5:52	34	5:43	51	5:30	68	4:59	
13	5:55	31	5:47	49	5:36	67	5:12	85	0:00	5
12	5:55	30	5:48	48	5:37	66	5:14	84	2:14	
11	5:56	29	5:48	47	5:38	65	5:16	83	2:58	
10	5:56	28	5:49	46	5:39	64	5:18	82	3:26	
9	5:56	27	5:49	45	5:39	63	5:20	81	3:45	
8	5:57	26	5:50	44	5:40	62	5:22	80	4:01	
7	5:57	25	5:50	43	5:41	61	5:23	79	4:13	
6	5:57	24	5:51	42	5:41	60	5:25	78	4:22	
5	5:58	23	5:51	41	5:42	59	5:26	77	4:30	
4	5:58	22	5:51	40	5:43	58	5:27	76	4:37	
3	5:58	21	5:52	39	5:43	57	5:29	75	4:43	
2	5:59	20	5:52	38	5:44	56	5:30	74	4:48	
1	5:59	19	5:53	37	5:44	55	5:31	73	4:53	
0	6:00	18	5:53	36	5:45	54	5:32	72	4:57	
		17	5:53	35	5:45	53	5:33	71	5:01	
		16	5:54	34	5:46	52	5:34	70	5:04	
		15	5:54	33	5:46	51	5:35	69	5:07	
		14	5:55	32	5:47	50	5:36	68	5:09	
14	5:56	32	5:49	50	5:40	68	5:20	86	0:00	4
13	5:56	31	5:50	49	5:41	67	5:22	85	2:27	
12	5:56	30	5:50	48	5:42	66	5:23	84	3:13	
11	5:56	29	5:51	47	5:42	65	5:25	83	3:41	
10	5:57	28	5:51	46	5:43	64	5:27	82	4:00	
9	5:57	27	5:51	45	5:43	63	5:28	81	4:15	
8	5:57	26	5:52	44	5:44	62	5:29	80	4:26	
7	5:58	25	5:52	43	5:45	61	5:31	79	4:35	
6	5:58	24	5:52	42	5:45	60	5:32	78	4:43	
5	5:58	23	5:53	41	5:46	59	5:33	77	4:49	
4	5:58	22	5:53	40	5:46	58	5:34	76	4:54	
3	5:59	21	5:53	39	5:47	57	5:35	75	4:59	
2	5:59	20	5:54	38	5:47	56	5:36	74	5:03	
1	5:59	19	5:54	37	5:47	55	5:37	73	5:07	
0	6:00	18	5:54	36	5:48	54	5:37	72	5:10	
		17	5:55	35	5:48	53	5:38	71	5:13	
		16	5:55	34	5:49	52	5:39	70	5:15	
		15	5:55	33	5:49	51	5:40	69	5:18	

TIMETABLE A--for stars in the **SAME CELESTIAL HEMISPHERE** as the circumpolar region.

Latitude	Dec	Time	Dec	Time	Dec	Time	Dec	Time	Dec	Time
3	87	12:00	69	6:31	51	6:14	33	6:07	15	6:03
	86	9:14	68	6:29	50	6:14	32	6:07	14	6:02
	85	8:27	67	6:28	49	6:13	31	6:07	13	6:02
	84	7:59	66	6:27	48	6:13	30	6:06	12	6:02
	83	7:41	65	6:25	47	6:12	29	6:06	11	6:02
	82	7:27	64	6:24	46	6:12	28	6:06	10	6:02
	81	7:17	63	6:23	45	6:12	27	6:06	9	6:01
	80	7:09	62	6:22	44	6:11	26	6:05	8	6:01
	79	7:02	61	6:21	43	6:11	25	6:05	7	6:01
	78	6:57	60	6:20	42	6:10	24	6:05	6	6:01
	77	6:52	59	6:20	41	6:10	23	6:05	5	6:01
	76	6:48	58	6:19	40	6:10	22	6:04	4	6:00
	75	6:45	57	6:18	39	6:09	21	6:04	3	6:00
	74	6:42	56	6:17	38	6:09	20	6:04	2	6:00
	73	6:39	55	6:17	37	6:09	19	6:04	1	6:00
	72	6:37	54	6:16	36	6:08	18	6:03	0	6:00
	71	6:35	53	6:15	35	6:08	17	6:03		
	70	6:33	52	6:15	34	6:08	16	6:03		
2	88	12:00	70	6:22	52	6:10	34	6:05	16	6:02
	87	8:47	69	6:20	51	6:09	33	6:05	15	6:02
	86	7:59	68	6:19	50	6:09	32	6:05	14	6:01
	85	7:34	67	6:18	49	6:09	31	6:04	13	6:01
	84	7:17	66	6:17	48	6:08	30	6:04	12	6:01
	83	7:06	65	6:17	47	6:08	29	6:04	11	6:01
	82	6:57	64	6:16	46	6:08	28	6:04	10	6:01
	81	6:50	63	6:15	45	6:08	27	6:04	9	6:01
	80	6:45	62	6:15	44	6:07	26	6:03	8	6:01
	79	6:41	61	6:14	43	6:07	25	6:03	7	6:00
	78	6:37	60	6:13	42	6:07	24	6:03	6	6:00
	77	6:34	59	6:13	41	6:06	23	6:03	5	6:00
	76	6:32	58	6:12	40	6:06	22	6:03	4	6:00
	75	6:29	57	6:12	39	6:06	21	6:03	3	6:00
	74	6:27	56	6:11	38	6:06	20	6:02	2	6:00
	73	6:26	55	6:11	37	6:06	19	6:02	1	6:00
	72	6:24	54	6:11	36	6:05	18	6:02	0	6:00
	71	6:23	53	6:10	35	6:05	17	6:02		
1	89	12:00	71	6:11	53	6:05	35	6:02	17	6:01
	88	7:59	70	6:10	52	6:05	34	6:02	16	6:01
	87	7:17	69	6:10	51	6:04	33	6:02	15	6:01
	86	6:57	68	6:09	50	6:04	32	6:02	14	6:00
	85	6:46	67	6:09	49	6:04	31	6:02	13	6:00
	84	6:38	66	6:08	48	6:04	30	6:02	12	6:00
	83	6:32	65	6:08	47	6:04	29	6:02	11	6:00
	82	6:28	64	6:08	46	6:04	28	6:02	10	6:00
	81	6:25	63	6:07	45	6:04	27	6:02	9	6:00
	80	6:22	62	6:07	44	6:03	26	6:01	8	6:00
	79	6:20	61	6:07	43	6:03	25	6:01	7	6:00
	78	6:18	60	6:06	42	6:03	24	6:01	6	6:00
	77	6:17	59	6:06	41	6:03	23	6:01	5	6:00
	76	6:16	58	6:06	40	6:03	22	6:01	4	6:00
	75	6:14	57	6:06	39	6:03	21	6:01	3	6:00
	74	6:13	56	6:05	38	6:03	20	6:01	2	6:00
	73	6:13	55	6:05	37	6:03	19	6:01	1	6:00
	72	6:12	54	6:05	36	6:02	18	6:01	0	6:00
0	90	0:00	0-89	6:00						

TIMETABLE B--for stars in the celestial hemisphere
OPPOSITE THE CIRCUMPOLAR REGION.

Dec	Time	Dec	Time	Dec	Time	Dec	Time	Dec	Time	Latitude
15	5:56	33	5:52	51	5:45	69	5:28	87	0:00	3
14	5:57	32	5:52	50	5:45	68	5:30	86	2:45	
13	5:57	31	5:52	49	5:46	67	5:31	85	3:32	
12	5:57	30	5:53	48	5:46	66	5:32	84	4:00	
11	5:57	29	5:53	47	5:47	65	5:34	83	4:18	
10	5:57	28	5:53	46	5:47	64	5:35	82	4:32	
9	5:58	27	5:53	45	5:47	63	5:36	81	4:42	
8	5:58	26	5:54	44	5:48	62	5:37	80	4:50	
7	5:58	25	5:54	43	5:48	61	5:38	79	4:57	
6	5:58	24	5:54	42	5:49	60	5:39	78	5:02	
5	5:58	23	5:54	41	5:49	59	5:39	77	5:07	
4	5:59	22	5:55	40	5:49	58	5:40	76	5:11	
3	5:59	21	5:55	39	5:50	57	5:41	75	5:14	
2	5:59	20	5:55	38	5:50	56	5:42	74	5:17	
1	5:59	19	5:55	37	5:50	55	5:42	73	5:20	
0	6:00	18	5:56	36	5:51	54	5:43	72	5:22	
		17	5:56	35	5:51	53	5:44	71	5:24	
		16	5:56	34	5:51	52	5:44	70	5:26	
16	5:57	34	5:54	52	5:49	70	5:37	88	0:00	2
15	5:57	33	5:54	51	5:50	69	5:39	87	3:12	
14	5:58	32	5:54	50	5:50	68	5:40	86	4:00	
13	5:58	31	5:55	49	5:50	67	5:41	85	4:25	
12	5:58	30	5:55	48	5:51	66	5:42	84	4:42	
11	5:58	29	5:55	47	5:51	65	5:42	83	4:53	
10	5:58	28	5:55	46	5:51	64	5:43	82	5:02	
9	5:58	27	5:55	45	5:51	63	5:44	81	5:09	
8	5:58	26	5:56	44	5:52	62	5:44	80	5:14	
7	5:59	25	5:56	43	5:52	61	5:45	79	5:18	
6	5:59	24	5:56	42	5:52	60	5:46	78	5:22	
5	5:59	23	5:56	41	5:53	59	5:46	77	5:25	
4	5:59	22	5:56	40	5:53	58	5:47	76	5:27	
3	5:59	21	5:56	39	5:53	57	5:47	75	5:30	
2	5:59	20	5:57	38	5:53	56	5:48	74	5:32	
1	5:59	19	5:57	37	5:53	55	5:48	73	5:33	
0	6:00	18	5:57	36	5:54	54	5:48	72	5:35	
		17	5:57	35	5:54	53	5:49	71	5:36	
17	5:58	35	5:57	53	5:54	71	5:48	89	0:00	1
16	5:58	34	5:57	52	5:54	70	5:49	88	4:00	
15	5:58	33	5:57	51	5:55	69	5:49	87	4:42	
14	5:59	32	5:57	50	5:55	68	5:50	86	5:02	
13	5:59	31	5:57	49	5:55	67	5:50	85	5:13	
12	5:59	30	5:57	48	5:55	66	5:51	84	5:21	
11	5:59	29	5:57	47	5:55	65	5:51	83	5:27	
10	5:59	28	5:57	46	5:55	64	5:51	82	5:31	
9	5:59	27	5:57	45	5:55	63	5:52	81	5:34	
8	5:59	26	5:58	44	5:56	62	5:52	80	5:37	
7	5:59	25	5:58	43	5:56	61	5:52	79	5:39	
6	5:59	24	5:58	42	5:56	60	5:53	78	5:41	
5	5:59	23	5:58	41	5:56	59	5:53	77	5:42	
4	5:59	22	5:58	40	5:56	58	5:53	76	5:43	
3	5:59	21	5:58	39	5:56	57	5:53	75	5:45	
2	5:59	20	5:58	38	5:56	56	5:54	74	5:46	
1	5:59	19	5:58	37	5:56	55	5:54	73	5:46	
0	6:00	18	5:58	36	5:57	54	5:54	72	5:47	
						0-89	6:00	90	0:00	0

The Problem with Clocks*

(*for those who wish to think about it!)

Since the sky and the earth's rotation are divided, each into 24 hours, we may think of them as two clocks. The clock which measures the earth's rotation, which we use to tell time, runs faster than the clock which measures the sky. This happens because the earth moves in its orbit around the sun while it makes its daily rotation.

When the clock measuring the earth's rotation has completed 24 hours, the clock measuring the sky has completed only about 23 hours and 56 minutes. When the clock measuring the sky has completed 24 hours, the clock measuring the earth's rotation has already completed about 24 hours and 4 minutes.

Fortunately, the planisphere (rotating star chart) takes this difference into account on a day-to-day basis. However, when the timetables are used with the planisphere to estimate starrise and starset on a given day, the fact that the clock measuring the earth's rotation moves faster than the clock measuring the sky is not taken into account. Of course, this adjustment adds only about 4 minutes or less to the time between rising and setting.

Star observers, who wish to compare their estimates with other calculations, may use the formula:

$$4\ (2\ T \div 24) = A,$$

where 4 is the approximate number of additional minutes possible, T is the time from the timetable (with the minutes converted to a decimal) and A is the additional number of minutes between rising and setting. We make the calculations for the example on pages 17-19:

$$4\ (3.16 \div 24) = .52 \text{ additional minutes (or 31 additional seconds).}$$

This time adjustment makes little difference when we round the observer's latitude and the star's location in the sky to the nearest degree; when we use a planisphere to estimate the time a star will be halfway between rising and setting; when the timepiece we use to plan observations is not set to the exact hour, minute and second; and when we do not make advanced calculations for local and atmospheric phenomena.

Now if clocks would tell us to enjoy the starry sky, and not to worry about clocks. . .

A Star Observer's Checklist

There are many atlases, road maps, planispheres, star maps and almanacs available to star observers. Chances are that you already have access to one or more of these items; so we didn't design our own.

This checklist mentions the items you will need; they include:

( ) 1. **A road map or an atlas** to determine the latitude of the observation site.

( ) 2. **A star map** to determine the star's declination ("latitude" in the sky).

( ) 3. **Any rotating star chart**, often called a planisphere, to estimate the time when a star will be halfway between rising and setting.

( ) 4. The timetables and explanations in this **Starrise and Starset Guide**, naturally!

( ) 5. **An almanac** (optional), so that the time of sunrise and sunset, the time of moonrise and moonset, as well as the time of twilight at dawn and dusk could be taken into account in your plans to observe the stars.

Looking at the Back Cover

A map of the area of the sky in the photograph on the front cover is printed on the back cover. (The photograph contains more stars than the map. The light of the fainter stars is responsible for the milky appearance of the photograph.) To compare the map to the photograph, look at the front cover and the back cover at the same time.

a page for notes

Approximate Declinations of Some Stars in 1990:

N 89.2°	Polaris	N 12.0°	Regulus	S 26.4°	Antares
N 46.0	Capella	N 08.8	Altair	S 29.7	Fomalhaut
N 45.3	Deneb	N 07.4	Betegeuse	S 52.7	Canopus
N 38.8	Vega	N 05.3	Procyon	S 57.3	Achernar
N 31.9	Castor	S 00.3	Delta Orionis	S 59.6	Beta Crucis
N 28.1	Pollux	S 08.2	Rigel	S 60.3	Beta Centauri
N 19.2	Arcturus	S 11.1	Spica	S 60.8	Alpha Centauri
N 16.5	Aldebaran	S 16.7	Sirius	S 63.0	Alpha Crucis

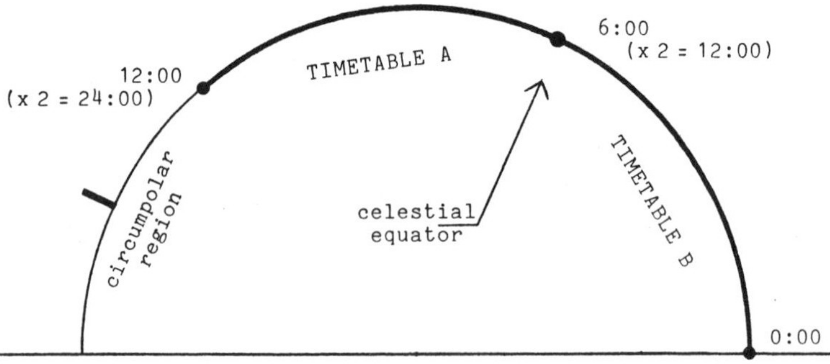